# Beyond Compliance:
## Leveraging AS9100D and ISO 9001:2015 Audit Findings for Continual Improvement

# Beyond Compliance:
## Leveraging AS9100D and ISO 9001:2015 Audit Findings for Continual Improvement

**Thomas Kannattumadom**

400 Commonwealth Drive
Warrendale, PA 15096-0001 USA
E-mail: CustomerService@sae.org
Phone: 877-606-7323 (inside USA and Canada)
724-776-4970 (outside USA)
Fax: 724-776-0790



**Library of Congress Catalog Number 2025939865**
**http://dx.doi.org/10.4271/9781468609677**



**ISBN-Print 978-1-4686-0966-0**
**ISBN-PDF 978-1-4686-0967-7**
**ISBN-epub 978-1-4686-0968-4**

**To purchase bulk quantities, please contact:** SAE Customer Service

E-mail: CustomerService@sae.org
Phone: 877-606-7323 (inside USA and Canada)
724-776-4970 (outside USA)
Fax: 724-776-0790

**Visit the SAE International Bookstore at books.sae.org**

**Publisher**
Sherry Dickinson Nigam

**Product Manager**
Amanda Zeidan

**Production and Manufacturing Associate**
Michelle Silberman

*To my valued clients in the aeronautical industry, in India and across the globe: The faith you've shown and the standards you champion continue to be a profound inspiration for my work.*

# Contents

## Chapter 10 - Nonconformities in Improvement Processes



## Chapter 11 - Action for Indicated Opportunities for Improvement

# Preface

"Implement AS9100 D/EN 9100:2018 for Business Excellence" (https://www.riverpublishers.com/book_details.php?book_id=1427) is a complete guide for implementing the modern quality management system (QMS) for aviation, space, and defense organizations. Audit and certification by a third-party certification body using the standard as the audit criteria demonstrate the organization's capability to its customers for its business scope. Any good value-added audit may bring out findings—nonconformities, gaps, and aberrations—in the practice of the QMS, which are to be addressed, and corrective actions are to be implemented effectively to restore the capability and excellence of the organization.

This book, *Beyond Compliance: Leveraging AS9100D and ISO 9001:2015 Audit Findings for Continual Improvement*, is a companion book for managers who must promptly complete corrective actions for every audit finding.

If you go through the IAQG-OASIS website, you will notice that there are 25,000 certified aviation, space, and defense organizations, and each one undergoes an annual audit. Specifically, there are 21,457 certifications for AS9100:2016 (also known as EN9100:2018, JISQ9100:2016, and KSQ 9100:2018 South Korean) globally.

According to the International Organization for Standardization (ISO), more than one million organizations worldwide are certified to ISO 9001, making it the most widely used QMS standard globally. In 2021, there were 1,077,884 ISO 9001 certificates issued worldwide. In 2022, China led the world with approximately 552,000 valid ISO 9001 certificates, followed by Italy and India. These numbers highlight the widespread adoption of ISO 9001:2015 across various industries and countries.

Every organization undergoes audits every year. Every audit uncovers nonconformities, which the organization must close after implementing corrective actions and verifying their effectiveness within a stipulated timeframe.

This practical guide is crafted in a simple storytelling style, designed to resonate with every executive in the workplace. Stories have a universal appeal, touching everyone involved and fostering practical solutions for challenges that arise on the shop floor or in the workplace. By engaging with the story, you will find it sparks ideas and inspires solutions that address real-world issues effectively.

I would like to call it "old wine in new bottle." The standard practices are old, benchmarked, and time-tested, and enjoying the taste is the key to success.

The names of the organizations where the nonconformities were identified during audits have not been disclosed. Each organization faced one or more nonconformities, which are presented here in the form of stories. All of these were followed up to ensure the implementation of effective corrective actions.

You might find that similar nonconformities or opportunities for improvement could be identified in your organization by an auditor. By engaging with the scenarios in this book, you will quickly envision customized solutions tailored to your specific challenges. Act on these insights, and you are likely to achieve meaningful success.

Thomas Kannattumadom
August 01, 2025

# List of Acronyms

**AAM** - Acceptance Authority Media

**AAQG** - Americas Aerospace Quality Group

**APAQG** - Asia Pacific Aerospace Quality Group

**AQMS** - Aerospace Quality Management System

**ASD** - Aerospace, Space, and Defense

**ATC** - Air Traffic Control

**BOM** - Bill of Materials

**CAD** - Computer-Aided Design

**CBs** - Certification Bodies

**CCB** - Configuration Control Board

**CDRs** - Critical Design Reviews

**CMP** - Configuration Management Plan

**CMS** - Configuration Management System

**CNC** - Computer Numerically Controlled

**CRM** - Customer Relationship Management

**CS** - Case Studies

**CSA** - Configuration Status Accounting

**D&D** - Design and Development

**DFARS** - Defense Federal Acquisition Regulation Supplement

**EAQG** - Europe Aerospace Quality Group

**EASA** - European Union Aviation Safety Agency

**ERP** - Enterprise Resource Planning

**FAA** - Federal Aviation Administration

**FAI** - First Article Inspection

**FAIR** - First Article Inspection Report

**FAR** - Federal Acquisition Regulations

**FIFO** - First-In First-Out

**FMEA** - Failure Mode and Effects Analysis

**FOD** - Foreign Object Debris

**GM** - General Manager

**HR** - Human Resources

**IAQG** - International Aerospace Quality Group

**ICOT** - IAQG Certification Oversight Team

**ISO** - International Organization for Standardization

**KPIs** - Key Performance Indicators

**MD** - Managing Director

**MIL-STD** - Military Standard

**MRO** - Maintenance, Repair, and Overhaul

**NC** - Nonconformity

**NCR** - Nonconformity Report

**OASIS** - Online Aerospace Suppliers Information System

**OEMs** - Original Equipment Manufacturers

**OFI** - Opportunities for Improvement

**OTD** - On-Time Delivery

**PDCA** - Plan–Do–Check–Act

**PDR** - Preliminary Design Review

**PEAR** - Process Effectiveness Assessment Report

**PFMEA** - Process Failure Mode and Effects Analysis

**PIFs** - Project Initiation Files

**PLM** - Product Lifecycle Management

**PMP** - Project Management Plan

**POs** - Purchase Orders

**PPC** - Production Planning and Control

**PPE** - Personal Protective Equipment

**PPL** - Precision Processors Ltd.

**PPV** - Production Process Verification

**PS** - Project Slip

**PSRB** - Product Safety Review Board

**QCI** - Quality Council of India

**QMS** - Quality Management System

**RCA** - Root Cause Analysis

**RFQ** - Request for Quotation

**rpm** - Revolutions per Minute

**RPN** - Risk Priority Number

**SCAR** - Supplier Corrective Action Request

**SEA** - Southeast Asian

**SMART** - Specific, Measurable, Achievable, Relevant, and Time-Bound

**SOP** - Standard Operating Procedure

**SOW** - Statements of Work

Chapter 01

# Aerospace Quality System in Action

## How the Quality System Was Born

Imagine a background story that might have happened 40 years ago.

### The Great Quality Summit of 1985: Birth of AS9100

It was a brisk morning in April 1985 when the world's most revered quality pioneers gathered at the SAE Convention Hall in Detroit. The hall was abuzz with excitement, as this was no ordinary meeting. Icons like Dr. William Edwards Deming, Joseph Juran, Philip Crosby, Armand Feigenbaum, Genichi Taguchi, Shigeo Shingo, Kaoru Ishikawa, Walter Shewhart, and many others had convened (some of them virtually) to tackle a monumental task: to reimagine quality management principles for a rapidly advancing industrial world.

The goal was ambitious—to lay the groundwork for a universal QMS that would elevate global standards. This summit would also address the aerospace, space, and defense (ASD) industries' unique challenges, proposing a sector-specific standard that could cater to their critical safety, reliability, maintainability, and compliance with statutory and regulatory needs.

### A Debate of Titans

The conference began with Deming taking the stage. "Quality is not just about fixing errors; it is about transforming systems," he said, setting the tone for the discussions. Juran nodded in agreement but interjected, "Transformation must be built on clear principles that drive consistency, efficiency, and customer satisfaction." Crosby, always the pragmatist, added, "And we must remember: quality is free. Investing in prevention costs far less than correcting errors."

The discussion soon expanded, with contributions from Ishikawa emphasizing teamwork and Taguchi focusing on designing for quality from the outset. Shingo brought in his expertise on mistake-proofing processes, while Shewhart reintroduced the idea of statistical control as a foundation for continual improvement.

## The Core Quality Principles

After days of rigorous debate and collaboration, the gurus distilled the following **QMS principles**:

1. **Customer Focus:** Prioritizing customer requirements and exceeding their expectations. The primary goal of a QMS is to meet customer requirements and exceed their expectations. Organizations should understand customer needs, provide quality products and services, and aim for customer satisfaction and loyalty.
2. **Leadership:** Establishing vision and direction and fostering an environment where quality thrives. Strong leadership establishes unity of purpose, vision, and direction for an organization. Leaders create an environment where people are motivated to contribute to achieving organizational goals. Leadership provides resources and support to achieve objectives, define a clear mission and vision, and foster a culture of trust and collaboration.
3. **Engagement of People:** Engaging and involving employees at all levels is essential to maximize their skills and contributions. People are more committed when they feel valued and understand their role in the organization's success. Key actions are as follows:
    - Encourage employee involvement in decision-making.
    - Provide training and opportunities for development.
    - Recognize and reward contributions.
    - Leverage the full potential of every individual within the organization.
4. **Process Approach:** A systematic approach to managing processes helps achieve consistent and predictable results. Understanding interrelated processes and optimizing them as a system enhances efficiency and effectiveness. Key actions are as follows:
    - Define key processes and their interconnections.
    - Monitor and control inputs, activities, and outputs.
    - Continually improve processes for better performance.
5. **Continual Improvement:** A commitment to continual improvement is vital for maintaining performance, adapting to changes, and meeting new challenges. Improvement should be part of the organization's culture. Key actions are as follows:
    - Identify opportunities for improvement (OFI).
    - Set improvement objectives and implement initiatives.
    - Learn from mistakes and successes to innovate and grow.
6. **Evidence-Based Decision-Making:** Decisions based on data and evidence are more likely to lead to desired results. Analyzing accurate information ensures a better understanding and reduces uncertainty. Key actions are as follows:
    - Collect reliable and relevant data.
    - Analyze trends and patterns to inform decisions.
    - Use facts, not assumptions, to guide actions.

7. **Relationship Management:** Building strong relationships with stakeholders—such as suppliers, partners, and customers—ensures long-term success. Mutually beneficial relationships enhance cooperation and value creation. Key actions are as follows:
   - Identify key stakeholders and understand their needs.
   - Foster open communication and collaboration.
   - Establish trust and mutual respect in partnerships.

By integrating these principles into its QMS, an organization can consistently deliver value to its customers and stakeholders while driving operational excellence and continual improvement.

The **process approach** is the most important principle. Here is an example.

### Example Process Elements in an ASD Organization

The product design and development (D&D) process is a critical example of a process in an ASD organization, where quality and on-time delivery (OTD) are paramount. Here is how the elements of this process are structured:

1. **Process Inputs** include the resources, information, and requirements necessary to initiate and execute the process effectively:
   - Customer Requirements: Specifications, functional needs, and compliance standards (e.g., AS9100D, Federal Acquisition Regulations (FAR), Defense Federal Acquisition Regulation Supplement (DFARS)).
   - Regulatory and Industry Standards: For example, the Federal Aviation Administration (FAA), the European Union Aviation Safety Agency (EASA), and Military Standard (MIL-STD).
   - Historical Data: Lessons learned, design checklists, and previous design iterations.
   - Resources: Design tools (CAD software, simulation tools), skilled engineers, and materials for prototyping.
   - Budget and Timeline: Constraints on costs and schedules.
2. **Activities** (how the process is done) are the specific steps involved in the D&D process:
   - Requirement Analysis: Review customer and regulatory requirements; define measurable objectives for the design.
   - Concept Development: Brainstorm and evaluate design concepts using trade studies or simulations; select the best-fit concept based on feasibility, cost, and performance.
   - Preliminary Design: Develop initial 3D models and drawings; conduct risk assessments (e.g., failure mode and effects analysis (FMEA)).
   - Detailed Design: Finalize drawings and specifications; perform stress analysis and simulations for validation.
   - Prototyping and Testing: Create prototypes; perform functional, structural, and environmental testing to validate performance.
   - Design Review and Approval: Present findings to the configuration control board (CCB); approve the design for production after verification and validation.
   - Handover to Production: Release controlled design documents; ensure the configuration management process is followed.

3. **Process Outputs** are the results or deliverables of the process: Approved design documentation: detailed drawings, specifications, and bill of materials (BOM).
   - Validated Prototypes: Tested models that meet performance requirements.
   - Change Records: Controlled logs of any design modifications during development.
   - Final Design Package: Released design ready for production, including compliance documentation.
4. **Process Objectives and Measurements** define what the process aims to achieve, and metrics ensure alignment with desired quality and OTD performance (**Table 1.1**):
   - Objectives: Deliver a compliant, reliable, and cost-effective design; ensure designs meet customer requirements and regulatory standards; meet agreed timelines without compromising quality.
5. **Continual Improvements:**
   - Carry out Structured Reviews: Regular gate reviews (concept, preliminary design, detailed design, etc.).
   - Integrated Tools: Use of product lifecycle management (PLM) systems for real-time tracking.
   - Configuration Control: Strict adherence to configuration management procedures for changes in the design output.
   - Resource Optimization: Assigning tasks to the most qualified engineers using scheduling software.
   - Continual Improvement: Gather lessons learned and data-driven enhancements to the process.

**Table 1.1** Key performance indicators.

| Objective | Metric | Target | Measurements during the last six months | The process is effective or not | What action is taken if the process is not effective |
|---|---|---|---|---|---|
| Quality assurance | Number of design errors, post-release | ≤1 per 100 drawings | — | — | — |
| Compliance | Percent compliance with customer standards | 100% | — | — | — |
| OTD | Milestone completion rate | ≥95% within scheduled dates | — | — | — |
| Efficiency | Average cycle time for the D&D process | <4 weeks per design phase | — | — | — |
| Customer satisfaction | Customer feedback scores | ≥4.5/5 | — | — | — |

By carefully managing each input, activity, and output while monitoring objectives and metrics, the D&D process ensures high-quality deliverables and consistent on-time performance in an ASD organization.

### The PDCA Cycle and the Process Approach

Shewhart's **Plan–Do–Check–Act (PDCA)** cycle was re-endorsed as the cornerstone of continual improvement. Deming passionately advocated for expanding its application across all industries, emphasizing its role in reducing process variation and driving excellence.

The **process approach** emerged as a critical theme, with Ishikawa and Shingo highlighting that quality is the cumulative result of well-managed, interconnected processes. "We cannot achieve true quality by looking at processes in isolation," Ishikawa remarked.

Every organization is unique, and in fact, every business model needs to fully align with the QMS. Strictly adhering to the QMS standard enables the organization to perform as desired.

The standard document (AS9100D/EN 9100:2018) standardizes QMS requirements to the greatest extent possible and can be used at all levels of the supply chain by organizations around the world. Its use should result in improved quality, cost, and delivery performance through the reduction or elimination of organization-unique requirements, effective implementation of the QMS, and wider application of good practice.

## The Aerospace Sector-Specific Amendment

During the final sessions, the spotlight shifted to the ASD industries, which demanded unique attention due to their critical safety and reliability requirements. Feigenbaum emphasized that a new standard was needed, one that built on the robust foundation of **ISO 9001** but introduced sector-specific requirements.

Crosby famously quipped, "We're not just talking about 'zero defects' here. In aerospace, zero defects can mean the difference between life and catastrophe."

The panel agreed on adding critical amendments, including:

- **Safety Requirements:** Ensuring product and workplace safety at every stage.
- **Statutory and Regulatory Compliance:** Addressing legal and regulatory obligations specific to ASD industries.
- **Reliability and Maintainability:** Designing systems for consistent performance and ease of maintenance throughout their lifecycle.
- **Risk-Based Thinking:** Introducing proactive risk assessment as a key component of quality planning.

The group that bargained for a standalone aerospace standard came to be known as the International Aerospace Quality Group (IAQG) in SAE. When more and more organizations from all over the world joined the group, they further subgrouped into three sectors for administrative convenience: Americas (AAQG), Europe (EAQG), and Asia Pacific (APAQG). Audits of ASD organizations are overseen by the IAQG Certification Oversight Team (ICOT), and IAQG also maintains information related to the Online Aerospace Suppliers Information System (OASIS).

### The Birth of AS9100 (9001 + 99)

As discussions wrapped up, a visionary proposal emerged: a standalone standard specifically for the ASD industries, based on ISO 9001 but enhanced with 99 additional sector-specific requirements (safety, statutory and regulatory requirements, reliability, and maintainability). This new standard would be known as **AS9100 Rev D** (latest) in the Americas (AS), EN 9100:2018 in Europe (EN), and JISQ 9100:2018 in the Asia Pacific (JISQ) sectors.

Juran remarked, "This is not just an extension of 9001. It is a paradigm shift for industries where quality, safety, reliability, and maintainability are non-negotiable."

Taguchi added, "With this, we are not just creating standards; we are shaping the future of aviation, space, and defense industries and equipment."

### The Final Resolution

The summit concluded with unanimous agreement on the following points:

1. **Refined QMS Principles:** A universal foundation for global quality management.
2. **Sector-Specific Requirements:** Acknowledgement of the unique needs of industries like aerospace and defense.
3. **A New Standard:** Development of AS9100 as a standalone framework incorporating ISO 9001 and additional requirements for safety, reliability, and maintainability, of course, complying with the statutory and regulatory requirements of the country.

Deming closed the summit with his characteristic optimism:

> *Quality is a journey without an end. Today, we have charted a course that will guide industries, nations, and humanity itself toward a better, safer future.*

The SAE Convention Hall erupted in applause, marking a historic moment that would shape the world of quality management for decades to come.

You can have a detailed study of the standard AS9100D (latest revision) and guidance for implementing the standard from the book *"Implement AS 9100 D/EN 9100:2018 for Business Excellence"* Quality Management Systems - Requirements for Aviation, Space and Defence Organizations (Same as EN 9100: 2016; Includes ISO 9001:2015) by Thomas Kannattumadom, aerospace industry experienced auditor. The book is available from River Publishers, Alsbjergvej 10, 9260 Gistrup, Denmark, www.riverpublishers.com (https://www.riverpublishers.com/book_details.php?book_id=1427) and Amazon.

## Audit Findings: Nonconformities and Opportunities for Improvement

Audit findings are the outcomes of assessing audit evidence against specified audit criteria. For this audit, the criteria are derived from the ISO 9001:2015 and AS9100D/EN 9100:2018 standards. Only the relevant clause numbers from these standards are referenced; for detailed requirements, please refer to the official standards published by ISO and SAE IAQG.

Audit findings may indicate either conformity or nonconformity (NC). Additionally, auditors can highlight OFI where no NC exists, encouraging continual enhancement and fostering a culture of excellence.

The scope of audit findings can encompass a range of areas, including financial reporting, regulatory compliance, and the design or effectiveness of internal controls. These findings are typically categorized as follows:

- **Conformity or NC:** Indicating adherence to, or deviation from, the standards with objective evidence (documents or records verified during the audit) (NCs indicated are those already issued or likely to be issued if the NC was found by the auditor).
- **Observations:** Notable insights or oral statements made during the audit process.
- **OFI:** Valuable points for advancing organizational practices on the path toward excellence.

This classification helps organizations identify strengths, address weaknesses, and seize opportunities for growth and improvement.

I have used fictitious names for 57 representative organizations where NCs were found during audits, all based on actual experience though some nonessential details have been changed to protect identities. Audit NCs are against ISO 9001:2015/AS9100D/EN 9100:2018 criteria, and their process effectiveness assessment ratings were level 2, level 3, and level 4. Every organization had one or more NCs reported during the annual audits.

## Consequences of NC with Industry Standards

Auditors often issue NCs during audits, and while it might feel like a bump in the road, it is an opportunity in disguise. Ignoring these NCs, however, can lead to significant challenges for your organization and its products/services/employees. The risks are real, but the solutions are well within your grasp.

Let's explore what can happen if follow-up actions are neglected and why addressing these findings promptly is essential for your success:

1. Certification Risks:
   - Suspension or Revocation: Failing to address NCs on time can result in the suspension or loss of your certification. This impacts your credibility and might even block access to lucrative contracts.
   - Renewal Challenges: Future audits can highlight unresolved issues, making recertification difficult and jeopardizing your organization's standing.
2. Regulatory and Legal Implications:
   - Noncompliance Fines: When NCs touch on legal requirements, ignoring them could lead to fines, penalties, or even legal proceedings.
   - Increased Oversight: Regulatory bodies might impose stricter controls, leading to more frequent inspections and increased scrutiny.
3. Operational Impact:
   - Inefficiencies: Unresolved issues can lead to recurring errors and process inefficiencies, hampering daily operations.
   - Escalating Costs: Poorly managed processes result in wastage, rework, and unnecessary expenses.
4. Reputational Damage:
   - Loss of Trust: Stakeholders, clients, and partners could lose faith in your organization's ability to deliver quality and maintain compliance.

- Bad Publicity: High-profile noncompliance issues can tarnish your reputation and attract negative attention.

5. Employee Morale and Engagement:
   - Low Motivation: When management neglects identified issues, employees may feel their efforts are undervalued, leading to dissatisfaction and disengagement.
   - Audit Apathy: Future audits might be met with resistance if employees perceive them as ineffective or meaningless.
6. Missed OFI:
   - Stagnation: NCs often highlight areas ripe for improvement. Ignoring them means missing out on opportunities to optimize and innovate.
   - Falling Behind Competitors: Organizations that embrace continual improvement will inevitably outpace those that do not.
7. Recurrence of Issues:
   - Recurring Problems: Unaddressed NCs often resurface, compounding inefficiencies and operational risks.
   - Major Failures: Small, ignored issues can snowball into larger, costlier failures.
8. Increased Audit Costs:
   - Wider Audit Scope: Neglected NCs mean more thorough audits in the future, increasing time and costs.
   - Extra Fees: Certification bodies (CBs) may charge for follow-up audits to verify overdue corrective actions.

### The Bottom Line

Addressing audit NCs is not just about ticking boxes—it is about protecting your organization's reputation, fostering a culture of excellence, and ensuring long-term sustainability. Each NC is a chance to learn, grow, and do better.

### Lessons from the Field

What happens when organizations fail to take timely action on NCs? The stories in this guide are drawn from real-life experiences, offering insights into common mistakes, overlooked opportunities, and how to turn gaps into growth.

This is not just a cautionary tale—it is a roadmap for achieving excellence, winning hearts, and creating a culture of continual improvement.

## Quality Lessons through Stories

We have a habit of learning from our mistakes that happened in the past. Systematic actions help to prevent recurrence of the mistakes and help us continually improve our performance.

An audit is a systematic, independent, and impartial activity documenting past mistakes and aberrations in practice and revealing objective evidence. Taking corrective actions, not mere corrections, and rectifications bring out positive improvements. It is the value addition that organizations get from third-party audits.

Conformity is the fulfillment of a requirement. The requirements are given in the international standard. If there is any non-fulfillment, it is a finding in the audit and it is reported by the auditor, in an NC report (NCR) format. Every NC needs to be acted upon, which includes analyzing the root cause of the defect/deficiency/mistake and taking corrective actions in such a way that such defect/deviation shall not happen any time in the future.

Effectiveness is achieving the results as planned or measuring the process objectives (quality, delivery schedule, etc.) against set targets for a specific period. Appropriate actions are mandatory if the measured results are not meeting the targets. Appropriate action includes analyzing the root cause of the shortage/deficiency and taking corrective actions in such a way that such deficit/shortfall in performance shall not happen any time in the future.

In a product/service realization process, meeting requirements and achieving results is always when the product/service delivered is to the entire satisfaction of the customers and the product/service delivered is exactly as per customer requirements.

Conformity and effectiveness are the essence of QMS practices, and they are verified during all audits, internal audits, or external audits and reported in the Process Effectiveness Assessment Report (PEAR).

In every customer meeting, technical convention, or general gathering, the management of the organizations proclaims their commitment to quality:

> *Our people are passionately committed to quality, utilize state-of-the-art technology to test and ensure the equipment they deliver is safe for the operations, durable, reliable, and maintainable. We provide the highest quality, and we care for our customers.*

It is a bunch of claims and data. It does not stick. They seldom go to the shop floor, deal with day-to-day issues, and listen to the technicians, but read and hear about everything and manage. They do not reveal the perception of some of the customers to the concerned people, but protect them and manage them. In case there is a shortage of supply for the progress of production, they pinch parts from other serviceable equipment. They do not show positive actions to mitigate the potential problems for long-term solutions. They manage. Businesses are always handling certain kinds of risk; to manage the same, the managers must proactively think of the possibilities and potential problems and take actions in advance so that the potential problems or mistakes or non-availability of necessary resources do not happen. They need to periodically review the same for continued effectiveness (desired results are being achieved or not) and take appropriate action to achieve the results at the earliest.

During my training sessions for the ASD industry personnel, I always told two things to participants:

1. If the standard is correctly understood, it gives the correct solutions to all day-to-day issues that crop up on the shop floor, or the standard directs you to think systematically to take appropriate actions for sustained progress.
2. You will get your certificate sooner or later whenever you apply for it to any CB. Certification does not mean that the quality of the product and OTD performance of the organization will always be 100%. It only conveys confidence to customers that the certified organization can deliver the required product, meeting the quality and OTD performance. There are a lot of preconditions to be fulfilled to practically demonstrate this capability.

Then, the participants usually raised two points, which I had to address:

- Ours is an established industry that regularly supplies parts to multiple customers. We have established additional facilities, and we want to get certified for more business with ASD customers. Our initial certification was delayed due to various NCs raised during audits that remained without proper corrective actions.
- We are a certified organization, but why are we not getting the desired benefits like cost reduction, savings, competitive advantage, consistent quality and productivity improvements, etc.?

I felt that analyzing the common audit findings/NCs in such organizations would give exposure to the real weaknesses in the organizations.

Taking appropriate corrective actions will surely restore their capability and credibility to excel in the business. The basic requirement for this is to understand the requirements in the QMS standard and implement them rigorously in their processes as solutions.

## Why Are QMS Audit Findings Presented as Stories?

The requirements are understood easily by the overall study of the QMS standards—ISO 9001:2015 and AS9100D/EN 9100:218, which are the criteria for the audit findings discussed in this book. If the problems/gaps/aberrations are presented as stories, the executive clearly understands them, as the benefit of a story is much more than the assimilation of technical descriptions. The story grips the reader/audience. He/she gets hooked on a story because he/she immediately falls in love with the characters and relates himself/herself to the situation. If these stories address their needs and interests, in the practice of the international QMSs, it is going to resonate with them and benefit them and their organizations.

Rob Biesenbach tells us that a business story has the power to break down walls, build trust, and influence people to act. It helps us to get more of what we desire. Therefore, this book adapts the style he indicated—identifying the character, his goal, challenge, and resolution (*Unleash the Power of Storytelling*, Eastlawn Media, IL 60201).

The stories shared here are based on my experiences during 35 years of service to the Indian Air Force and 20 years of subsequent auditing and consulting around the world. While names and details have been changed to protect privacy, the lessons are universal. These insights are meant to empower and guide organizations toward better practices, stronger compliance, and sustained success.

## Guide to Case Study Numbering

Case study discussions of audit findings are grouped into a few major chapters by theme, in line with the SAE Standard, and numbered in sequence from CS 1 to CS 137. Each case study mentions the relevant clause from AS9100D. Some of these clauses also appear in ISO 9001:2015 and have been noted as such.

Specific details about the following sector-specific requirements are given in the case studies indicated against them (**Table 1.2**).

**Table 1.2** NCR and corrective action requirement.

| Serial no. | NCR and corrective action requirement | Serial number of relevant case studies (CS) |
|---|---|---|
| 1 | Critical items | 6, 56, 69, 80 |
| 2 | Key characteristics | 6, 56, 69, 80 |
| 3 | Special requirements | 6, 56, 69, 80 |
| 4 | Product safety (8.1.3) | 28, 40 |
| 5 | Awareness on QMS documents, product or service conformity, product safety, and ethical behavior (7.3 h) | 3, 7, 13, 17, 20, 28, 59, 66, 129 |
| 6 | Operational risk management (8.1.1) | 34, 35 |
| 7 | Configuration management (8.1.2) | 9, 37, 38, 40, 56 |
| 8 | Prevention of counterfeit parts (8.1.4) | 29, 30, 41, 63, 130 |
| 9 | Aerospace supplier management (8.4.1.1) | 60 |
| 10 | Control of equipment, tools, and software programs (8.5.1.1) | 87 |
| 11 | Validation and control of special processes (8.5.1.2) | 88 to 93 |
| 12 | Production process verification (8.5.1.3) | 95 |
| 13 | Organizational risk and opportunities (6.1) | 17, 18, 19 |
| 14 | Human factors/human error (10.2.1(b)/8.5.1(g)) | 15, 42, 69, 76, 77, 127 |
| 15 | Foreign object damage/debrief (8.1(a), 8.5.1(o), 8.5.4.(b)) | 99 |

Chapter 02

# Audit Nonconformities in Management Processes: Context of the Organization

## CS 1 [AS9100D Clauses 4.1, 4.2] Ignoring the Obvious: AeroTech's Wake-Up Call to Excellence

In the bustling manufacturing facility of AeroTech Systems, Management Representative Daniel Harper was known for his relentless commitment to excellence. However, he found himself in a precarious position when a third-party audit revealed a significant NC related to clauses 4.1 and 4.2 of AS9100D/EN 9100:2018. Appropriate evidence was recorded by the auditor below the NC in the NCR.

The findings were damning: the organization had failed to explain its context, neglected essential SWOT or PESTLE analyses to understand the market landscape, ignored customer and other stakeholders' feedback, and overlooked potential risks associated with new legislation or technological advancements. Worst of all, the needs and expectations of key stakeholders had been disregarded and not recorded for periodic review.

Daniel's credibility and the organization's certification were at stake. Determined to turn the tide, he convened an emergency meeting with his cross-functional team. The room buzzed with tension as he laid out the statement of NC and explained the gravity of the situation.

*Clauses 4.1 and 4.2 context not explained, SWOT analysis or PESTLE analysis not done, to understand the wider market landscape. Ignoring customer feedback and not adapting products or services accordingly. Failing to identify potential risks associated with new legislation or technological advancements. Not considering the needs and expectations of key stakeholders (customers, suppliers, employees).*

The auditor has recorded appropriate evidence below the NC.

"We need immediate containment action," Daniel began. "Our credibility is on the line, and so is our customers' trust. Let's address the symptoms now while we dig deeper into the root cause."

## Containment Action

The first step was to re-establish communication channels with key stakeholders. Daniel assigned his customer relations team to reach out to major clients and gather feedback on AeroTech's performance. Simultaneously, the market analysis team was tasked with performing a rapid SWOT and PESTLE analysis to identify gaps in understanding the external and internal landscape and issues that are likely to affect the organizational performance.

Daniel personally took charge of organizing a stakeholder meeting with suppliers and employees to discuss their concerns and expectations. Within days, the immediate actions bore fruit. Customer concerns were acknowledged, and preliminary insights into market trends were compiled, buying the organization some time to address the underlying issues.

## Root Cause Analysis (RCA)

With the containment measures in place, Daniel turned his focus to identifying the root causes of the NC. The team employed a structured approach, using tools like the 5 Whys and Fishbone Diagram to trace the problem to its origin.

## Key Findings

- **Lack of Structured Processes:** The organization had no formal process for performing SWOT or PESTLE analyses.
- **Neglecting Feedback Mechanisms:** Customer feedback was collected sporadically and not systematically analyzed or acted upon.
- **Reactive Risk Management:** The risk management framework was outdated, focusing on immediate operational risks rather than long-term strategic risks.
- **Siloed Operations:** Departments worked in isolation, resulting in a fragmented understanding of stakeholder needs and expectations.

## Corrective Actions

Armed with these insights, Daniel and his team designed a robust corrective action plan.

1. Institutionalizing SWOT and PESTLE Analyses:
   - A dedicated strategy team was established to conduct periodic SWOT and PESTLE analyses.
   - Training sessions were held to ensure all team members understood how to apply these tools effectively.
2. Enhancing Feedback Systems:
   - A customer feedback platform was launched to collect, categorize, and analyze feedback systematically.
   - A biweekly review meeting was instituted to track customer feedback and integrate it into decision-making.

3. Proactive Risk Management:
    - The risk management framework was overhauled to include emerging risks, such as new legislation and technological advancements.
    - A risk register was created and reviewed quarterly to ensure ongoing relevance. The risk register also included external and internal issues that had the potential to affect organizational performance.
4. Breaking Down Silos:
    - Cross-functional teams were formed to improve communication and collaboration among departments.
    - Stakeholder expectations were mapped and aligned with organizational objectives, ensuring no group was overlooked.

### Results

Six months later, a follow-up audit was conducted. The auditor, impressed with the transformations, noted significant improvements in AeroTech's approach to strategic planning and stakeholder management. Customer satisfaction scores had risen by 25% and delivery timelines improved by 15% as the organization became more attuned to market and customer needs.

Daniel's leadership had not only restored the credibility of AeroTech's certification but also positioned the organization as a benchmark for quality and adaptability. As he reflected on the journey, he realized the importance of embedding continual improvement into the organization's DNA. The lesson was clear: understanding the context and addressing stakeholder needs were not just checkboxes for certification but essential drivers of long-term success of the organization.

## CS 2 [AS9100D Clauses 4.1, 4.2 with 6.1] Recovering from Risky Ignorance

In the bustling manufacturing facility of AeroTech Systems, Management Representative Daniel Harper was known for his relentless commitment to excellence. However, he found himself in a precarious position when a third-party audit revealed a significant NC related to clauses 4.1, 4.2, and 6.1 of AS9100D/EN 9100:2018.

*Clauses 4.1 and 4.2 may be read with 6.1 risks and opportunities arising from external and internal issues and needs and expectations of interested parties are not identified and managed through the risk/opportunity register.*

Appropriate points of evidence were recorded by the auditor below the NC.

The findings were incriminating: the organization had failed to identify and manage risks and opportunities arising from external and internal issues, as well as the needs and expectations of interested parties, through its risk/opportunity register.

Daniel's credibility and the organization's certification were at stake. Determined to turn the tide, he convened an emergency meeting with his cross-functional team. The room buzzed with tension as he laid out the statement of NC and explained the gravity of the situation.

"We need immediate containment action," Daniel began. "Our credibility is unprotected, and so is our customers' trust. Let's address the symptoms now while we dig deeper into the root cause."

### Containment Action

The first step was to create a temporary risk/opportunity register. Daniel assigned a task force to identify and document risks and opportunities associated with ongoing projects, pending customer orders, and upcoming regulatory changes. Simultaneously, key stakeholders were contacted to gather their immediate concerns and expectations, ensuring no critical issues were overlooked.

Within a week, the containment action provided a preliminary framework for risk and opportunity management. This temporary register highlighted urgent risks, such as supplier delays and changing customer requirements, enabling the organization to implement interim controls.

### RCA

With the containment measures in place, Daniel turned his focus to identifying the root causes of the NC. Using a structured problem-solving approach, the team employed tools like 5 Whys and Failure Mode and Effects Analysis (FMEA).

### Key Findings

- **Inadequate Framework:** The organization lacked a formalized process to capture and analyze external and internal issues.
- **Overlooked Expectations:** Stakeholder expectations were not systematically reviewed or integrated into strategic planning.
- **Reactive Risk Management:** The existing risk management process was limited to operational concerns, ignoring strategic risks and opportunities.
- **Communication Gaps:** Departments operated in silos, hindering the flow of critical information about risks and opportunities.

### Corrective Actions

Armed with these insights, Daniel and his team developed a robust corrective action plan.

1. Establishing a Comprehensive Risk/Opportunity Register:
   - A centralized risk and opportunity register was created to document and track all identified risks and opportunities.
   - A dedicated team was assigned to ensure the register was updated regularly and aligned with organizational objectives.
2. Integrating Stakeholder Expectations:
   - A formal process was established to gather and analyze the needs and expectations of customers, suppliers, employees, and regulatory bodies.
   - Regular stakeholder review meetings were scheduled to ensure alignment with organizational goals.
3. Enhancing Risk Management Processes:
   - The risk management framework was expanded to include strategic risks, such as market trends, technological advancements, and regulatory changes.
   - Training sessions were conducted to ensure all employees understood their roles in identifying and mitigating risks.

4. Improving Cross-Functional Communication:
    - Cross-functional teams were formed to break down silos and promote collaboration.
    - A shared digital platform was implemented to facilitate real-time communication and updates on risks and opportunities.

### Results

Six months later, a follow-up audit was conducted. The auditor, impressed with the transformations, noted significant improvements in AeroTech's risk and opportunity management processes. The organization's ability to anticipate and address risks had markedly improved, resulting in more reliable delivery timelines and enhanced customer satisfaction.

Daniel's leadership had not only restored the credibility of AeroTech's certification but also strengthened its capability to deliver conforming products on time. Reflecting on the journey, Daniel realized the importance of proactive risk management and stakeholder engagement as pillars of sustainable success. The lesson was clear: identifying and managing risks and opportunities was not just about compliance—it was about building a resilient and adaptable organization.

## CS 3 [AS9100D Clause 4.3] Lost in Scope: How AeroNova Found Its Focus

The warm sunlight streamed through the blinds as Robert Shaw, the Management Representative at AeroNova Systems, sipped his coffee and prepared for the day ahead. Known for his unflinching commitment to quality, Robert had faced his fair share of challenges in ensuring AeroNova's adherence to the stringent requirements of AS9100D/EN 9100:2018. Today, he was about to tackle one of the most critical issues yet: an NC in defining the organization's scope.

### The Problem Emerges

*The audit report sat heavily on his desk, marked with the unmistakable bold text: "Statement of NC: the scope is not clearly defined." Appropriate evidence was recorded by the auditor below the NC. This finding aligned with clause 4.3 of the AS9100D standard, which mandated a precise definition of the organization's scope. Without this, the very foundation of AeroNova's QMS was at risk.*

Robert read the statement again, determined to uncover how this oversight could have occurred. The scope statement, intended to clearly define AeroNova's boundaries activities and relevance to customer requirements, had become muddled over time. Without a clear scope definition, their certification—and by extension, their credibility—was on shaky ground.

### Immediate Containment Action

The first step was containment. Robert convened an emergency meeting with key stakeholders, including the quality team, operations leaders, and senior management.

"The audit finding is a serious concern," Robert began, his tone measured but firm. "We need immediate action to prevent further confusion. For now, all internal and external communications referencing our scope will be paused. A temporary clarification will be issued to address any urgent customer inquiries."

The team agreed on a provisional statement that outlined the organization's activities—design, manufacturing, and maintenance of aerospace components—to ensure customers and auditors had a clear understanding in the interim.

### RCA

Containment addressed the surface issue, but Robert knew that real improvement required digging deeper. He formed a cross-functional team to perform an RCA. Using tools like 5 Whys and Ishikawa diagrams, the team uncovered multiple contributing factors:

- **Ambiguity in Leadership Input:** Over the years, leadership changes introduced varying interpretations of the organization's activities.
- **Outdated Documentation:** The scope statement had not been reviewed or updated in alignment with recent expansions in service offerings.
- **Lack of Awareness:** Some employees were unfamiliar with the importance of a clearly defined scope in maintaining QMS integrity.

Robert summarized the findings in a clear report and presented them to the senior leadership team. "This is not just about correcting the scope," he explained. "It is about ensuring our systems and processes prevent this kind of issue from recurring."

### Corrective Action: A Permanent Solution

With leadership's full support, Robert led the charge to implement a corrective action plan. The following measures were prioritized:

1. **Revising the Scope Statement:** The team drafted a new, comprehensive scope statement, explicitly detailing AeroNova's activities and their relevance to customers.
    - Example: "Manufacturing and supply of precision aerospace components, specializing in high-performance assemblies for commercial and defense applications."
    - NA Clauses: 8.3: D&D justified as it is not a requirement from the customers, and the organization has not yet built the capability to design aerospace components. It is a build-to-print company.
2. **Leadership Alignment Workshops:** Robert facilitated sessions with senior leaders to ensure a unified understanding of the organization's scope and its implications for customer satisfaction and certification credibility.
3. **Regular Reviews and Updates:** A review process was established to revisit the scope statement annually or whenever significant changes occurred in the organization's activities.
4. **Training and Awareness Campaigns:** A company-wide program was launched to educate employees on the importance of the scope in maintaining a robust QMS.
5. **Integration into QMS Audits:** The revised scope statement was embedded into internal audit checklists to verify alignment across all processes.

### The Results

Months later, a follow-up audit was conducted. This time, the auditor's comments were glowing.

"The scope statement is exemplary," the lead auditor said during the closing meeting. "It not only meets the requirements of clause 4.3 but also provides clarity and confidence in AeroNova's capabilities."

Internally, the changes had ripple effects. Teams were more aligned, processes were more streamlined, and customer confidence soared as OTD rates improved. AeroNova's QMS now stood as a robust framework for delivering conforming products to customers—on time, every time.

### A Legacy of Continual Improvement

As Robert reflected on the journey, he felt a deep sense of accomplishment. Addressing the NC had not only restored the credibility of AeroNova's certification but had also strengthened the organization's capability to deliver on its promises.

"Conformance is not just about meeting standards," he told his team during a celebratory meeting. "It is about building trust with our customers and ensuring we're always ready to adapt and improve. We must write what we do and do what we are capable of."

Robert knew that challenges would continue to arise, but with a culture of proactive problem-solving and a commitment to excellence, AeroNova was ready for the future.

## CS 4 [AS9100D Clause 4.4.1] The Apex Challenge: Process Improvement Story

Jacob Alden, the Management Representative, sat at his desk, reviewing the NCR with a sinking feeling.

*The audit's findings were clear: "Processes are not explained, and process charts are often mistaken as organizational department charts."*

Appropriate evidence was recorded by the auditor below the NC. This issue, flagged under criteria 4.4.1 of AS9100D/EN 9100:2018, jeopardized the organization's certification and its ability to consistently deliver conforming products on time.

The stakes were high, but Jacob was determined to turn this setback into an opportunity. His goal was twofold: restore the organization's credibility and enhance its process capability to better meet customer expectations.

### Immediate Containment

Jacob convened an emergency meeting with his team the next morning. "This is not just a documentation problem," he began. "It is a systemic issue that affects how we operate. We need to act swiftly to contain the impact."

The team agreed on immediate containment actions:

- Manual Recall: All distributed copies of the Apex Manual were recalled.
- Temporary Addendum: A supplementary document was issued to clarify the distinction between processes and organizational departments.

- Customer Communication: Key customers were informed about the issue, along with assurances of corrective measures to maintain delivery schedules.

### RCA

With containment in place, Jacob initiated an RCA. A cross-functional team was assembled, and they used the 5 Whys methodology to trace the problem's origins. After hours of rigorous discussion and investigation, the root causes emerged:

- Inadequate Training: Employees lacked knowledge of process mapping techniques and standards.
- Legacy Charts: Outdated organizational charts were repurposed as process charts, leading to confusion.
- Miscommunication: Poor coordination among departments during the manual's preparation.

Jacob summarized the findings. "This is a systemic failure, not just an oversight. We need to overhaul our approach to process documentation."

### Corrective Action

To address the root causes, the team devised a comprehensive corrective action plan:

1. Redefine Process Documentation Standards: A dedicated task force was formed to create clear templates for process maps. These templates highlighted inputs, outputs, responsibilities, and interactions. Visual aids like swim lane diagrams were introduced to differentiate processes from organizational structures.
2. Training Programs: Workshops on process mapping and AS9100 standards were conducted for all relevant employees. Jacob personally led sessions to underline the importance of accurate documentation.
3. Independent Review: An external consultant was engaged to review the revised Apex Manual before its release, ensuring compliance with AS9100 standards and best practices.

### Results and Reflection

Three months later, the revised Apex Manual was finalized and distributed. It featured well-defined processes, clear flowcharts, and a distinct separation from organizational charts. Employees praised the manual's usability, and customers noted improved efficiency in product delivery.

During the follow-up audit, the external auditors commended the organization. "This is a remarkable turnaround," one remarked. "Not only have you resolved the NC, but you've also set a benchmark for process clarity."

Jacob felt a surge of pride as he heard the feedback. The crisis had been a wake-up call, but it had also catalyzed meaningful change. The organization's credibility was restored, and its capability to deliver conforming products had improved significantly.

That evening, as Jacob packed up his office, he reflected on the journey. Challenges like these, he thought, reveal weaknesses but also provide opportunities to rebuild stronger. Picking up his journal, he wrote: *"Adversity is the best teacher. Today, we proved that resilience and teamwork can turn setbacks into success stories."*

## CS 5 [AS9100D Clause 4.4.1] The Crane Conundrum: Oversight to Overhead

The clang of metal and the rhythmic hum of machinery echoed through the factory floor as the managing director (MD) led the auditor, clipboard in hand, through the bustling small-scale Japanese industry. The Indian auditor's sharp eyes scanned every detail, dissecting the processes and scrutinizing compliance. This audit, critical for AS9100D certification, demanded perfection.

Reaching the broaching machine section, the auditor stopped abruptly. His gaze fixed on an overhead crane slowly hoisting a gleaming heavy turbine blade used in the gas turbine of generators. The movement was steady but unmistakably essential for the production process.

"Overhead cranes," the auditor began, his voice firm but calm, "play a critical role in safety and compliance. Could I see the safety certification records?"

The maintenance manager shuffled uncomfortably, his eyes darting to the MD who requested the auditor to sit in an adjoining room. From the maintenance manager, a flurry of papers emerged, and the auditor reviewed the crane safety certifications one by one. Then, the auditor's pen paused mid-air over a particular document.

"This certificate," he said, holding it up, "is dated four years ago."

The room fell silent. The MD, a composed figure, turned toward the manager, eyebrows raised in question. The manager cleared his throat and hesitated before replying, "Sir, that crane was unused for the past four years. We only resumed its operation last month."

The auditor's expression hardened. "Has it undergone the statutory safety certification before being brought back into use? There is such a system in my country."

The manager fumbled, and a translator was called to convey the auditor's question. The response came back as a simple, uneasy, "Yes, Japan has a system for safety certifying overhead cranes. But... it wasn't done in this case."

The MD's calm demeanor cracked. "This is serious. What does this mean for our certification process?"

The auditor's tone was measured but stern. "This constitutes an NC under clause 4.4.1 of AS9100D. Your QMS does not address the requirements of safety certification for overhead cranes used in production. This is a direct breach of compliance and poses significant safety risks." Appropriate evidence, such as the serial number of the crane certified four years back, was also recorded below the NC.

The MD frowned; his voice now edged with urgency. "What's the containment action? We must continue production!"

The auditor's reply was unequivocal. "Stop using the crane temporarily until it is safety certified. Production must proceed without compromising safety standards. Arrange alternative methods or equipment if necessary."

A tense silence followed, broken only by the hum of the factory floor. The MD turned to his quality department head. "Immediate action. Ensure alternative equipment is arranged for loading and unloading. Expedite the crane's safety certification."

The quality team mobilized on the spot, logging the NC into the system. They initiated containment, an RCA, and a corrective action plan.

*The auditor documented his findings meticulously: statement of NC: clause 4.4.1 of AS9100D—the organization's QMS does not address the requirements of safety certification of overhead cranes used for production.*

Appropriate evidence was recorded by the auditor below the NC.

As the auditor departed to continue his audits, the MD glanced at the manager. "This is a lesson we cannot afford to repeat. Compliance is non-negotiable, not just for certification but for safety and integrity."

The factory's usual rhythm resumed, but the air carried a charged tension—a sharp reminder of the cost of oversight in a world where every detail matters.

## CS 6 [AS9100D Clause 4.4.1] A Wake-Up Call: Aligning Documentation with Standards

It was a crucial day for AeroSoft Solutions, a leading software organization specializing in engineering services for ASD clients. The recertification audit for their QMS was underway, and the lead auditor, Richard Blake, was meticulously reviewing the organization's documentation. He has seen some statements in the quality manual such as "They declared and documented that they did comply with the requirements of 'Monitoring and Measuring Resources' (7.1.5) and 'Validation and Control of Special Processes' (8.5.1.2)."

While the auditor was visiting the site before the document verification, none of the processes could demonstrate the use of any measuring equipment for their services. The quality chief confirmed that they do not have any special processes to do validation. It was a software/engineering services organization. "The organization could have documented that those clauses do not apply to their business/QMS." The auditor made a mental note. The tension in the room was palpable as Blake's sharp eyes scanned through the quality manual, project management plan (PMP), and related compliance records.

"Your organization's reputation precedes you," Blake began, his tone both measured and firm, "but your QMS documentation does not align with the high standards you're known for." He placed the documents on the table and looked directly at the team, composed of AeroSoft's quality manager, project leads, and compliance officers.

Blake picked up a particular section of the PMP and read aloud: "Determination of requirements:

identify statutory/regulatory requirements, special requirements, key characteristics, critical items, value additions, or other requirements not stated by the customer." He paused for effect, then continued, "And the response here is: 'No special requirements. No NAA/FAA/EASA/DGCA requirements. No key characteristics. Nil. None. Not Available. Not Applicable. Customer will provide.'" He set the paper down and folded his hands.

"This approach is unacceptable," Blake stated firmly. "Your organization cannot absolve its responsibilities with such generic statements. Critical characteristics and requirements must be proactively identified to add value and ensure customer satisfaction. If the customer has not provided specific details, you must seek clarification and document the findings properly. This is not just a matter of ticking boxes; it is about maintaining the integrity of your operations."

The room was silent, save for the faint hum of the air conditioning. AeroSoft's quality manager, Priya Menon, finally spoke. "Blake, we acknowledge your concerns. However, in some cases, our customers provide incomplete statements of work, leaving us in a difficult position. We've relied on their inputs to avoid overstepping."

Blake nodded thoughtfully. "I understand the challenges, Menon, but let me remind you of a fundamental principle. 'Practice whatever you document, document whatever you practice.' While this precept has guided industries for decades, it is no longer sufficient. Today, the focus must shift to: 'Whatever you document or practice shall be as per the standard.' Standards exist to ensure consistency, compliance, and quality. They are the foundation of trust in sectors like aerospace."

He continued, "Take your PMP as an example. It is an excellent guideline for operational planning and control, but its effectiveness is undermined by the vague and dismissive responses in your compliance records. This undermines your credibility and places your organization at risk. A strong QMS should explicitly address all requirements—statutory, regulatory, and customer-specific—even if it means initiating a dialogue with the customer to clarify ambiguities."

The audit team's faces reflected a mix of acknowledgement and resolve. Blake's words were a wake-up call.

"Here is my suggestion," Blake said, pulling out a marked-up version of the PMP. "Review your documentation with precision. Identify what you have and, more importantly, what you do not have. Address each requirement of the standard in your quality manual explicitly. Vague answers such as 'Not Applicable' or 'Customer will provide' should only be used when supported by evidence or correspondence from the customer. A proactive approach will not only improve compliance but also strengthen your relationship with your customers."

By the end of the audit session, the AeroSoft team had a clear path forward. They resolved to overhaul their documentation process, ensuring it was both comprehensive and aligned with standards. Menon summed up the team's determination: "This is not just about passing an audit. It is about living up to our reputation and delivering excellence in every project."

As Blake signed off his audit notes, he left them with a final thought: "Standards are not constraints; they are enablers. They guide us to practice and document in ways that elevate our work. Embrace them, and you'll not only meet expectations but exceed them."

## CS 7 [AS9100D Clause 4.4.1] Certification Crossroads: A Manager's Statutory Compliance

The hum of machinery filled the sprawling aerospace shop floor as Vikram, the shop manager, wiped the sweat off his brow. He'd been racing against time for months, pushing his team to implement a QMS to meet AS9100D standards before the impending certification audit.

"If we nail this, it opens doors to bigger contracts," he kept telling himself. But the haste had left little room to dive deep into the intricate requirements of the standard.

### The Audit Day

The morning of the audit arrived with an air of nervous anticipation. The lead auditor, Iyer, a sharp-eyed industry veteran, meticulously inspected the documentation and toured the facility. His questions were precise, cutting through the surface.

By afternoon, the verdict came.

*The organization's QMS does not document, identify, or address the requirements for customers and applicable statutory and regulatory QMSs as required by clause 4.4.1 of AS9100D. The auditor has noted appropriate evidence for the finding.*

Vikram's heart sank. "What do we do now?" he asked, feeling the weight of failure.

Iyer, however, wasn't there to just point fingers. "Let's use this as a stepping stone. Address the finding through containment actions, conduct an RCA, and establish permanent corrective measures. I'll return to verify the effectiveness of your actions."

### The Comeback Plan

1. Containment Action:
   - Vikram halted all processes tied to noncompliant areas.
   - A task force was set up to review immediate gaps and relay customer and statutory requirements to relevant teams.
2. RCA: The team carried out a 5 Whys analysis:
   - Why did we miss requirements? → Rushed implementation.
   - Why was it rushed? → Pressure to meet deadlines.
   - Why was there not a detailed checklist? → Lack of training.
   - Why no training? → Assumption that basic awareness sufficed.
   - Why the assumption? → No dedicated quality compliance officer.
3. Corrective Action:
   - Appoint a compliance officer.
   - Conduct AS9100D training for all relevant personnel.

- Revise and document QMS to comprehensively include customer and applicable statutory and regulatory requirements.
- Implement an internal audit schedule to ensure ongoing compliance.

### The Return Audit

Weeks later, Iyer returned. He meticulously reviewed the actions taken, scanned updated documentation, and interviewed the team.

After a thorough evaluation, he smiled. "I'm impressed. You've not only corrected the issue but have strengthened your overall quality system. This will lead to improved customer satisfaction, fewer defects, and a competitive edge in the market."

Vikram sighed with relief, the tension melting away. The certificate soon followed, not just as a piece of paper but as a testament to their resilience and commitment to quality.

Their journey to certification had started with a stumble, but they had risen stronger, ready to soar.

## CS 8 [AS9100D Clause 4.4.2] Mind the Gap: A Journey of Trust and Leadership

John, the Management Representative for a manufacturing company, was eager to ensure that the organization's QMS passed its first audit phase smoothly. After months of preparation, he was confident that all the necessary documentation, including the quality manual, was in perfect order.

The day of the audit arrived. The external auditor, Sarah, meticulously reviewed the documents provided. Everything seemed to align with the ISO 9001/AS9100D requirements—until she reached the quality manual.

"John," Sarah said, her tone professional but firm, "your quality manual lacks a critical component. It does not document the sequence and interaction of the processes within your organization. Is there any?"

John was momentarily taken aback. The oversight seemed minor but was significant enough to warrant an NC. The auditor may not report the stage 1 findings as NCR but requires positive improvement in the documentation which the auditor insisted. "I'll address it right away," he assured her, determined to resolve the issue promptly.

Back at his desk, John reviewed the guidelines and collaborated with his team to create a comprehensive process map. The map visually outlined the sequence of each process—from order receipt to delivery—and illustrated how these processes interacted to ensure consistent quality and customer satisfaction.

A revised quality manual, now featuring the new process maps, was submitted to Sarah for review. She studied it carefully, nodding in approval.

"This is excellent work," Sarah said. "The process maps clearly define the sequence and interaction of the processes. It ensures clarity and alignment with the standard's requirements."

With the NC resolved, Sarah marked the phase 1 audit as successful, allowing the company to proceed to the next stage.

John felt a sense of relief and pride. This experience reinforced the importance of meticulous attention to detail in maintaining a robust QMS. More importantly, it was a lesson in continual improvement—an ethos that underpinned their organization's quality journey.

## CS 9 [AS9100D Clause 4.4.2] Chaos to Certification: Compliance to Mandatory Documents and Records

The continual, metallic clattering and grinding noise, often described as a mix of clanging, banging, and whirring sounds, filled the air as Rajan Menon, the Shop Manager at Aerovista Aerospace Solutions, paced the production floor. His brows furrowed in concentration, he clutched a clipboard brimming with notes and action. The AS9100D certification audit was just a week away, and Rajan had been driving the team relentlessly to implement the QMS. But in the rush, details had blurred.

The Stage 1 audit day arrived, bringing with it Aryan Das, the lead auditor from a reputed CB. With a calm demeanor and an eagle eye for compliance, Aryan began his systematic evaluation.

### The Audit Findings

As the audit progressed, Aryan's questions grew more pointed.

"Can you show me the documented information about your relevant interested parties?" he asked.

Rajan flipped through a binder and presented a brief list. Aryan's pen tapped on the table. "This is a start, but it lacks a general description. It doesn't show how these parties influence your QMS."

Next, Aryan examined the QMS scope. "You've mentioned the general scope, but boundaries and applicability are vague. For instance, how do you define the exclusions?"

The most glaring issue surfaced when Aryan requested to see process documentation. The binder contained only compliance statements.

"There's no detailed description of your processes, their interactions, or sequence," Aryan noted, his voice steady but firm. "And the assignment of responsibilities is listed as duties, not as integrated authority over process outcomes."

After wrapping up the audit, Aryan summarized his findings:

*The organization's QMS does not maintain complete documented information of the requirements as per clause 4.4.2 of AS9100D. You must establish proper containment actions, investigate the root causes, and implement and document the permanent corrective actions before we can proceed to Stage 2.*

## RCA and Corrective Action

Rajan convened a crisis meeting. Using the 5 Whys technique, the team drilled down to the root cause:

1. Why did documentation fail?—Rushed implementation.
2. Why the rush?—Pressure to meet audit deadlines.
3. Why no detailed process descriptions?—Lack of understanding of the standard's requirements.
4. Why lack of understanding?—No formal AS9100D training.
5. Why no training?—Assumed prior ISO 9001 knowledge would suffice.

Armed with this insight, Rajan initiated corrective actions:

- **Containment Action:** Halt ongoing implementation and prioritize critical documentation.
- **Training:** Conduct AS9100D workshops for key personnel.
- **Documentation Overhaul:** Create detailed process maps, define interactions, and assign process ownership.
- **Verification:** Conduct internal audits to ensure documentation completeness and accuracy.

## Stage 2 Audit and the Turnaround

When Aryan returned for the Stage 2 audit, the transformation was evident. Processes were meticulously mapped, boundaries clearly defined, and roles explicitly assigned.

The list of mandatory documents and records as noted below specially prepared by the management representative for establishing the QMS was worth emulating by everyone:

| | |
|---|---|
| Context of the organization that includes external and internal issues and a general description of relevant interested parties | Clause 4.2 a |
| Scope | Clause 4.3 |
| Operation of its processes including processes, scope, description of processes, sequence and interactions, responsibilities, and authorities (quality manual) | Clause 4.4.2 |
| Quality policy | Clause 5.2 |
| Objectives | Clause 6.1 |
| Identify documented information for effective QMS | As Required |
| Control of documented information including records | Clause 7.5.3.1 |
| Quality plan maintained | Clause 8.1 |
| Configuration management maintained | Clause 8.1.2 |
| D&D requirements are met | Clause 8.3.2 |
| Production records, product details, drawings, specifications | Clause 8.5.1 |
| NC control process | Clause 8.7.1 |
| NC and corrective action management process | Clause 10.1 |

The following were noted as the mandatory documented information (retained as records) as per Standard 9100 Rev D:

| | |
|---|---|
| Operation of its processes completed as planned | Clause 4.4.2 |
| Fitness of purpose M&M resources, calibration reports | Clause 7.1.5.1, 7.1.5.2 |
| Competence | Clause 7.2 |
| Production operation, planning, and control | Clause 8.1 |
| Review of requirements | Clause 8.3.2 |
| Design inputs | Clause 8.3.3 |
| Design review, verification, validation | Clause 8.3.4 |
| Design outputs | Clause 8.3.5 |
| Design changes | Clause 8.3.6 |
| Supplier evaluations | Clause 8.4.1 |
| Supplier records, retention periods | Clause 8.4.3 |
| Inspection records | Clause 8.5.1.2 |
| Validation of special processes | Clause 8.5.1.2 |
| Production process verification, (PPV) (first article inspection, (FAI)) | Clause 8.5.1.3 |
| Traceability | Clause 8.5.2 |
| Customer property | Clause 8.5.3 |
| Control of changes to production | Clause 8.5.6 |
| Release of products | Clause 8.6 |
| Control NC and results | Clause 8.7.2, 9.1.1 |
| Audit and audit results | Clause 9.2 |
| Results of management reviews | Clause 9.3.3 |
| NC and results of corrective action | Clause 10.2 |

After completing the verification, Aryan smiled. "I must commend your team. The corrective actions were well executed, and the documentation is now robust."

He highlighted the organization's overall gains:

- Process Clarity: Seamless process interaction and ownership.
- Employee Empowerment: Clear responsibilities foster accountability.
- Quality Culture: Enhanced understanding of quality standards.
- Customer Confidence: Strengthened reputation in the aerospace sector.

With the AS9100D certification secured, Rajan leaned back in his chair, finally breathing easy. The journey had been intense, but the lessons learned were invaluable. The organization emerged not just certified, but stronger and more resilient.

And that, Rajan thought with a smile, was worth every late night and tough audit finding.

Chapter 03

# Audit Nonconformities in Management Processes—Leadership

## CS 10 [AS9100D Clause 5.1] Leadership in Action: Turning Oversight into Opportunity

Ramesh, the MD of a small-scale Indian manufacturing company, stood in his office, gripping the NCR tightly.

*The audit findings under AS9100D/EN 9100:2018 delivered a sharp critique: "Even though the processes are managed to the best of his ability, commitment is not clear to the leadership." The auditor recorded appropriate evidence below the NC.*

Ramesh's mind raced. The organization's reputation was at stake, as was its certification. Despite pushing hard to meet delivery schedules, a few product rejections had crept in—an outcome he could not ignore any longer. His goal was clear: safeguard the company's certification, maintain OTDs, and restore faith in his leadership.

### Immediate Containment

That evening, Ramesh called an emergency meeting with his Management Representative, Ananya Mehta, and department heads.

"We need immediate action," Ramesh began, his voice resolute. "We cannot afford any more product rejections or questions about our leadership. Ananya, I need you to lead the containment efforts."

The team proposed swift measures:

- **Customer Communication:** An apology letter and assurance of corrective actions were sent to customers who had experienced delays or quality issues.
- **Temporary Measures:** Production schedules were adjusted, introducing additional quality checks at critical stages to minimize rejections.

- **Internal Messaging:** Ramesh issued a company-wide statement, reinforcing the importance of quality and timely delivery.

### RCA

With the immediate crisis contained, Ananya initiated a structured RCA. She convened a cross-functional team to delve into the issue, using tools like 5 Whys and Pareto analysis.

After days of investigation, the team identified key contributors to the NC:

1. **Lack of Leadership Visibility:** Employees perceived Ramesh's focus on pushing deliveries as neglecting quality and process adherence.
2. **Insufficient Engagement:** Leadership was not consistently involved in quality reviews or improvement initiatives.
3. **Reactive Approach:** The organization's efforts often addressed symptoms rather than underlying systemic issues.

Ananya summarized the findings to Ramesh. "The processes are running, but without visible and consistent leadership commitment, the team lacks a clear sense of direction."

### Corrective Action

Ramesh, recognizing the depth of the issue, worked closely with Ananya to develop a robust corrective action plan:

1. **Leadership Training and Engagement:** Ramesh enrolled in leadership training focused on quality management and employee engagement. He also instituted a monthly "quality forum," where he personally reviewed performance metrics and led discussions on improvement initiatives. He monitored product and service conformity and OTD performance.
2. **Policy and Process Overhaul:** The quality policy was rewritten to emphasize leadership commitment to both delivery and conformance. Processes were revised to include mandatory leadership reviews at key stages.
3. **Employee Feedback Mechanism:** An anonymous feedback system was introduced, allowing employees to share concerns about leadership and processes. Ramesh committed to addressing these issues transparently.
4. **Recognition and Accountability:** A "quality champion" program was launched to recognize employees who excelled in process adherence. At the same time, department heads were held accountable for ensuring alignment with the updated quality policy.

### The Turnaround

Three months later, the next audit provided a stark contrast to the previous one. The auditors noted the changes, particularly Ramesh's active involvement in quality management. "This is a commendable transformation," the lead auditor remarked. "The leadership's commitment is now evident across all levels of the organization."

Employees also noticed the difference. Ramesh's presence in quality reviews and his acknowledgement of their efforts fostered a renewed sense of pride. Product rejections decreased significantly, and OTDs improved.

### Reflection

That evening, Ramesh sat in his office, reflecting on the journey. The NC had been a hard lesson, but it had also taught him the true essence of leadership: being present, engaged, and committed.

He wrote in his journal: "Leadership is not just about driving results; it is about inspiring trust and setting the right example. Today, we proved that even setbacks could pave the way for growth."

## CS 11 [AS9100D Clause 5.1.2] Promises to Performance: A Leadership Quest

Ramesh, the MD of a small-scale Indian manufacturing firm, leaned back in his chair, the weight of the audit report pressing heavily on his mind.

*The statement of NC under criteria 5.1.2 of AS9100D/EN 9100:2018 was a direct challenge to his leadership: "Customer focus is not only honoring/respecting customers but also delivering 100% quality and OTD. Data showed that performance was not as per expectations during the last three years. Customer complaints were discussed and promised to be solved, but gaps remained."*

The auditor recorded appropriate evidence below the NC indicating the percentage product conformity and OTD performance.

For Ramesh, this was not just a technicality. It threatened the company's certification, reputation, and long-term relationships with clients. His goal was clear: maintain the certification, restore customer trust, and improve performance while continuing to meet delivery schedules, even if it meant addressing painful truths.

### Immediate Containment

Ramesh convened an urgent meeting with his Management Representative, Ananya Mehta, and the leadership team.

"We've always focused on delivery," Ramesh began, "but this report makes it clear: focusing on delivery alone is not enough. We need to address quality and customer expectations head-on."

To contain the immediate fallout, the team implemented several actions:

- **Customer Communication:** A formal letter was sent to key customers, acknowledging recent shortcomings and outlining immediate steps being taken to address their concerns.
- **Enhanced Inspection:** Additional quality checks were introduced for ongoing projects, ensuring that only conforming products were shipped.
- **Expedited Complaint Resolution:** A dedicated team was tasked with resolving outstanding customer complaints within a set timeframe.

### RCA

With the immediate containment actions underway, Ananya led a structured RCA. Using the 5 Whys method and reviewing performance data from the last three years, the team uncovered several underlying issues:

1. **Reactive Approach:** The company's focus on immediate delivery often overshadowed proactive quality management.
2. **Data Overlooked:** Performance metrics were collected but not analyzed effectively, leading to missed OFI.
3. **Inconsistent Customer Feedback Handling:** Complaints were acknowledged but not always resolved systematically.

Ananya summarized her findings. "The root of the problem is not just operational—it is cultural. We need to make quality and customer focus core values, not afterthoughts."

## Corrective Action

Ramesh worked closely with Ananya to develop a robust corrective action plan that would permanently address the NC:

1. **Customer Focus Policy Overhaul:** The company's customer focus policy was revised to emphasize 100% quality and OTD as non-negotiable priorities. This policy was communicated across all levels of the organization.
2. **Data-Driven Decision-Making:** A performance dashboard was introduced to track metrics such as OTD, defect rates, and customer complaints. Weekly reviews ensured accountability and transparency.
3. **Customer Engagement Program:** Regular meetings with key customers were scheduled to gather feedback, discuss expectations, and review ongoing improvements. A "customer delight" initiative was launched to reward teams that excelled in meeting customer needs.
4. **Training and Awareness:** Workshops were conducted to instill a culture of quality and customer focus. Employees were trained in problem-solving techniques and the importance of data analysis.

## The Turnaround

Over the next six months, the impact of these changes became evident. The performance dashboard showed a steady decline in defect rates and an improvement in OTD. Customer complaints dropped significantly, and feedback from key clients highlighted renewed confidence in the company.

During the follow-up audit, the lead auditor noted the transformation. "Your commitment to customer focus is now evident in both your processes and results," he remarked. "This is a remarkable turnaround."

Employees felt the difference too. Ramesh's involvement in weekly reviews and his openness to feedback inspired a new sense of accountability and pride. The "customer delight" initiative became a driving force for innovation and excellence.

## Reflection

That evening, as Ramesh reviewed the latest performance metrics, he felt a deep sense of satisfaction. The journey had been challenging, but it had also reaffirmed the values he wanted his company to stand for.

He wrote in his journal: "Customer focus is not a checkbox exercise—it is a commitment. Today, we proved that by listening, learning, and leading with intent, we can turn promises into performance."

## CS 12 [AS9100D Clause 5.2.1] A Journey to Certification: The Story of AeroFab Solutions

AeroFab Solutions, a fledgling ASD start-up in Bangalore, India, had recently entered the competitive industry with ambitious plans. Their primary focus was "built-to-print" manufacturing, offering precision engineering services to global aerospace giants. As a young organization, AeroFab Solutions was keen to establish itself as a reliable partner by achieving AS9100D certification—an essential benchmark for QMSs in aerospace.

### The Audit Begins

The company eagerly welcomed auditors for the initial certification (phase 1) of AS9100D. It was a significant milestone that would showcase their commitment to quality and open doors to bigger business opportunities.

The audit team meticulously reviewed AeroFab Solutions' policies, processes, and documentation.

### The NC

*During the review, they identified an NC related to clause 5.2.1 of AS9100D, which deals with the establishment of quality policy.*

The auditor noted the following issues with AeroFab Solutions' quality policy statement:

- The policy statement merely reiterated the company's scope of work—"built-to-print manufacturing"—without reflecting broader organizational objectives.
- The statement lacked alignment with strategic goals, such as customer satisfaction, continual improvement, and compliance with industry standards.
- The policy was verbose and unclear, making it difficult for stakeholders to comprehend and relate.

The auditor documented the NC and explained it to the Management Representative, Raghav Kumar.

### The Follow-Up Action

Determined to address the NC promptly, Raghav led the organization in revisiting the quality policy. He gathered inputs from key stakeholders, including the leadership team, engineering heads, and shop-floor supervisors. The objective was to create a policy statement that was concise, meaningful, and aligned with the organization's vision.

After deliberations and brainstorming, AeroFab Solutions revised its quality policy to include the following elements:

- A commitment to meeting customer requirements and enhancing satisfaction.
- A focus on continual improvement of processes and adherence to industry standards.
- Alignment with the strategic objective of becoming a preferred aerospace partner globally.

The revised policy read: "AeroFab Solutions is committed to delivering high-quality, built-to-print aerospace components that meet customer requirements and exceed expectations. We prioritize innovation, process improvement, and adherence to global standards to establish ourselves as a trusted industry leader."

To ensure organization-wide understanding, Raghav conducted training sessions for employees to explain the new policy and its significance. The updated policy was prominently displayed in the facility and communicated to all stakeholders.

### Verification and Acceptance

During the follow-up audit, the auditor reviewed the revised quality policy and the corrective actions taken. Impressed by the concise yet impactful statement, the auditor noted that the policy now effectively reflected the organization's objectives and was well aligned with AS9100D requirements.

The NC was closed, and AeroFab Solutions successfully passed phase 1 of the certification audit.

### Conclusion

This experience marked a turning point for AeroFab Solutions. The team not only learned the importance of aligning policies with organizational goals but also embraced the spirit of continual improvement. As the company moved closer to full certification, it carried forward the lessons learned, solidifying its position as a rising star in the aerospace industry.

The journey of AeroFab Solutions serves as a testament to the resilience and determination of start-ups striving for excellence in the competitive ASD sector.

## CS 13 [AS9100D Clause 5.2.1] The Tale of the Unaware Organization

During the annual ISO 9001 audit at Acme Enterprises, the external auditor, Collins, roamed the premises, meticulously observing operations and engaging with employees to assess their understanding of the company's QMS.

It was during one such interaction with a junior technician, Ravi, in the assembly section, that Collins stumbled upon a significant NC.

"Can you tell me what the quality policy of your organization is?" asked Collins, expecting at least a broad understanding of the policy.

Ravi hesitated, scratching his head. "Uh... quality policy? I'm not sure, sir. I just focus on completing my tasks as instructed."

Concerned, Collins continued his investigation by questioning several other employees across departments—warehouse staff, office clerks, and even some supervisors. To his dismay, most employees were either unaware of the quality policy or had no idea how their roles contributed to achieving the organization's quality objectives.

In his closing meeting with the management team, Collins presented his findings.

*NC: clause 5.2 of AS9100D—awareness of the quality policy. A significant portion of the workforce is unaware of the quality policy and its role in contributing to its objectives. This is a critical issue and requires immediate attention.*

The Management Representative, Ms. Anika, took the observation very seriously. Determined to resolve the NC, she spearheaded a comprehensive corrective action plan:

1. **Awareness Training Sessions:** Anika organized mandatory training sessions for all employees. These sessions were engaging, using real-life examples to explain the quality policy and how each department and role contributed to its success.
2. **Quality Policy Displays:** She ensured the quality policy was prominently displayed in common areas like the cafeteria, main hallways, and near workstations. Posters included simplified diagrams linking roles to quality objectives.
3. **Departmental Meetings:** Department heads were tasked with holding weekly discussions with their teams to reinforce the quality policy and address any questions.
4. **Interactive Quizzes and Activities:** To make learning fun and engaging, employees participated in quizzes and workshops. Incentives, such as small rewards, were given to top performers, further motivating the workforce.
5. **Follow-Up Evaluation:** Anika implemented a system where supervisors routinely checked their team's understanding of the quality policy during team meetings and documented the outcomes.

After three months of intensive effort, Collins returned for a follow-up audit. This time, the results were markedly different.

In the manufacturing unit, Ravi confidently explained, "Our quality policy is about delivering products that meet customer satisfaction. My role is crucial because the parts I assemble contribute directly to the reliability of our products."

In the warehouse, a storekeeper proudly shared, "By ensuring proper inventory management, I'm helping to reduce errors and delays, which aligns with our quality objectives."

Collins noted the transformation in his final report: "NC closure: clause 5.2. The organization has successfully addressed the NC by implementing a robust awareness program. Employees now demonstrate a clear understanding of the quality policy and their contribution to its achievement."

The management team breathed a sigh of relief, but the incident was a wake-up call. From then on, Acme Enterprises ensured that awareness of the quality policy became an integral part of its culture.

The experience improved compliance and boosted employee morale and engagement, proving that quality is not just a policy—it is a collective commitment.

## CS 14 [AS9100D Clause 5.2.2] Lost in Translation: Turning Policies into Practice

In the heart of an Indian small-scale industrial corporation, General Manager (GM) Arvind Sharma prided himself on his leadership. His goal was straightforward: impress the auditors, maintain the organization's AS9100 certification, and ensure the company's continued ability to deliver products to customers, even if a few complaints surfaced now and then. Arvind believed that every complaint was addressed, albeit without a systematic approach to permanent corrective actions. However, during the audit, an unsettling observation shook his confidence.

### The NC

The auditor's statement of NC under criteria 5.2.2 of AS9100D/EN 9100:2018 was clear:

> *There is no evidence that everybody in the organization is aware of the policy and objectives or that they are contributing toward achieving them. Some supervisors could not state the policy or objectives, citing that the policy was in English and had not been explained to them. It appeared to be displayed only for auditors and customers.*

The auditor recorded appropriate evidence below the NC.

Arvind immediately recognized the gravity of the issue. It was not just about certification; it was about ensuring alignment and commitment across the organization.

### Immediate Containment

Determined to address the NC, Arvind called an emergency meeting with Khadgi Desai and the department heads.

"We need immediate action," Arvind began. "The policy and objectives should not just be words on a board. They must be understood, embraced, and acted upon by everyone."

To contain the issue, the team implemented the following measures:

- **Supervisor Training:** A crash course was conducted for all supervisors, explaining the quality policy and objectives in simple terms and their relevance to day-to-day tasks.
- **Translated Materials:** The policy was translated into Hindi and the regional language for better comprehension.
- **Visible Commitment:** Posters with the translated policy and simplified objectives were prominently displayed in work areas.

### RCA

Khadgi led an RCA to understand why the gap had occurred. Using tools like the 5 Whys method, the team uncovered the following issues:

1. **Policy Language Barrier:** The English-only policy was inaccessible to many employees, especially shop-floor workers.
2. **Lack of Training:** New employees and supervisors were not trained on the policy and objectives during onboarding or routine operations.
3. **Symbolic Display:** The policy's display was treated as a formality, not as a tool for engagement or guidance.

Khadgi summarized, "Our policy is not connecting with the people it is meant to guide. This is a systemic issue, not just a one-off oversight."

### Corrective Action

Arvind and Khadgi developed a corrective action plan to address the root causes and ensure sustained compliance:

1. Simplified Policy Communication:
   - The policy was rewritten in a concise, relatable manner.
   - Key objectives were broken down into measurable tasks linked to each department's responsibilities.
2. Multilingual Support:
   - Policies and objectives were translated into regional languages.
   - Audiovisual presentations were created to explain the policy engagingly.
3. Employee Training Program:
   - Regular training sessions were instituted, focusing on the policy, objectives, and each employee's role in achieving them.
   - A quiz-based assessment was introduced to ensure understanding.
4. Leadership Engagement:
   - Managers and supervisors were tasked with conducting monthly meetings to review how their teams were contributing to organizational objectives.
5. Monitoring and Feedback:
   - An internal audit mechanism was set up to periodically verify awareness and alignment with the policy.
   - Feedback loops were established to capture employee suggestions for improvement.

### The Turnaround

Over the next few months, the transformation was evident. Employees on the shop floor could confidently articulate the organization's policy and objectives, linking them to their daily tasks. Supervisors displayed a newfound sense of ownership and pride in their roles. The organization's improved alignment was reflected in better quality metrics and fewer customer complaints. During the follow-up audit, the auditor noted the changes. "Your team's awareness and engagement with the policy is impressive," he remarked. "This is not just compliance; it is commitment."

### Reflection

That evening, as Arvind reviewed the day's events, he felt a mix of relief and pride. The journey had not only secured the certification but also strengthened the organization from within. He jotted down a note to himself: "A policy is not a decoration for auditors; it is a compass for every individual in the organization. Today, we turned that belief into reality."

## CS 15 [AS9100D Clause 5.3] The Missing Job Descriptions: Flexibility vs Formality

Rajiv Prasad found himself in a precarious position during the first surveillance audit. The auditor had flagged a significant NC under criteria 5.3 of AS9100D/EN 9100:2018:

*Organizational roles, responsibilities, and authorities are not clearly defined for some personnel. Verbal instructions were given to supervisors and some managers to complete tasks without any documentation support.*

The auditor recorded appropriate evidence below the NC.

This revelation pointed to a deeper issue in the organization's operational framework. GM Ramesh Devendran Murthy, known for his practical approach, believed in keeping things flexible. His philosophy was simple: the best person should do the job, and everyone in the division should be competent to handle any role. However, this informal approach had now become a liability.

### The Immediate Containment Action

As soon as the NC was identified, Rajiv convened an emergency meeting with Ramesh and the human resources (HR) team. They decided on the following containment actions:

- **Task Allocation Clarification:** Verbal instructions were documented for the ongoing projects to ensure clarity.
- **Appointing Management Representative:** The head of the organization appointed the management representative, indicating his additional duties, responsibilities, and authorities in addition to the existing ones.
- **Temporary Role Definitions:** Interim role and responsibility charts were created for all supervisors and managers.
- **Auditor Communication:** Rajiv assured the auditor that a structured approach to addressing the issue was underway.

While these steps provided immediate relief, Rajiv knew they were merely stopgaps. A permanent solution was needed to align with AS9100D standards.

### RCA

Rajiv and the HR team conducted an RCA to identify why the NC had occurred. They used the Fishbone Diagram approach and identified the following causes:

1. **Absence of Formal Documentation:** Roles, responsibilities, and authorities were not documented, relying instead on GM's verbal directives.
2. **Over-Reliance on Informality:** The culture of flexibility and multitasking led to ambiguous task ownership.

3. **Lack of Awareness:** Managers and supervisors were not aware of the importance of formally defining roles to meet certification requirements.

### Corrective Action Plan

With the root causes identified, Rajiv worked closely with HR to implement a corrective action plan:

1. Defining Roles and Responsibilities:
   - A comprehensive organizational structure was developed, clearly defining roles, responsibilities, and authorities for all personnel.
   - Job descriptions were created and communicated to employees.
2. Documentation Framework:
   - A policy document was drafted to outline the procedure for defining and updating roles and responsibilities.
   - Supervisors and managers were provided with templates to document task assignments and authorities.
3. Training and Awareness:
   - A series of workshops were conducted to educate employees about the importance of clearly defined roles and their connection to organizational objectives.
   - Managers were trained on how to document and communicate task allocations effectively.
4. Monitoring and Review:
   - A periodic review process was implemented to ensure that role definitions remained relevant and updated.
   - Internal audits were scheduled to verify compliance with the new system.

### The Outcome

Three months later, during a follow-up audit, the organization showcased its revamped system. The auditor was impressed by the clear documentation and alignment of roles and responsibilities with organizational goals. Supervisors and managers demonstrated a thorough understanding of their tasks, supported by well-defined job descriptions.

Rajiv reflected on the transformation. "We've turned a challenge into an opportunity," he noted. "This process has not only resolved the NC but also strengthened our operations."

### Lessons Learned

That evening, as Ramesh Devendran Murthy reviewed the day's success with Rajiv, he admitted, "I've always believed in flexibility, but I see now that structure and clarity are equally important. From now on, we'll balance both approaches."

The journey had been a learning experience for everyone involved, reinforcing the idea that clear roles and responsibilities are foundational to organizational success and compliance.

## CS 16 [AS9100D Clause 5.3] From Verbal to Perfect: A Wake-Up Call

During the first surveillance audit of a small-scale ASD industry in India, Management Representative Prakash Desai faced a significant setback. The auditor flagged an NC under criteria 5.3 of AS9100D/EN 9100:2018 and recorded appropriate evidence below the NC:

> *There was no evidence to show that the management was asking every process owner to follow the process approach, asking to measure the objectives, and ensuring that the processes were delivering their intended outputs. Process objectives were not established and measured for processes like purchase, quality, and planning departments.*

Prakash had been operating on verbal instructions from GM Arun Verma, who believed in a results-oriented approach without delving into formalities like defining objectives or documenting processes. While this philosophy might have worked in the short term, it left critical gaps in compliance and organizational effectiveness.

### Immediate Containment Action

Realizing the gravity of the NC, Prakash called for an urgent meeting with department heads and process owners. Together, they implemented the following containment measures:

- **Temporary Objectives:** Interim objectives were defined for the purchase, quality, and planning processes.
- **Process Documentation:** Process flows and key performance indicators (KPIs) were outlined and communicated to the respective team.
- **Auditor Communication:** Prakash assured the auditor that a structured corrective action plan was being developed to address the issue comprehensively.

These immediate steps stabilized the situation, but Prakash knew a permanent solution was essential to prevent recurrence.

### RCA

Prakash assembled a cross-functional team to conduct an RCA. Using the 5 Whys technique, they uncovered the following issues:

1. **Lack of Awareness:** The GM and process owners were not fully aware of the process approach requirement under AS9100D.
2. **Absence of Structured Objectives:** No framework existed for setting, monitoring, or evaluating objectives for individual processes.
3. **Informal Culture:** Verbal instructions were preferred over documented processes, leading to inconsistencies.
4. **Limited Accountability:** Process owners were not held accountable for ensuring their processes delivered intended outputs.

### Corrective Action Plan

With the root causes identified, Prakash collaborated with the team to implement a corrective action plan:

1. Process Approach Training:
    - A series of workshops were conducted to train management and process owners on the process approach, emphasizing the importance of establishing objectives and measuring outputs.
2. Defining Process Objectives:
    - Specific, measurable, achievable, relevant, and time-bound (SMART) objectives were established for all key processes, including purchase, quality, and planning.
    - All process owners were requested to make their own process maps indicating all the elements of the process. How and when to measure the process objectives by the process owner was explained in a circular addressed to all process owners.
    - These objectives were linked to ultimate organizational goals and customer requirements.
3. Process Monitoring and Measurement:
    - KPIs were defined for each process, and regular periodic monitoring mechanisms were established.
    - Monthly review meetings were scheduled to evaluate process performance and address any deviations.
4. Accountability Framework:
    - Roles and responsibilities were clearly defined for all process owners and their support staff.
    - A system of periodic internal audits was introduced to ensure adherence to the process approach.

## The Turnaround

Three months later, during a follow-up audit, the organization showcased its revamped process management system. The auditor observed that:

- All processes had clearly defined objectives aligned with organizational goals.
- Process owners demonstrated a thorough understanding of their responsibilities and the KPIs associated with their processes.
- Regular monitoring and review mechanisms were in place, ensuring processes delivered their intended outputs.
- Whenever there is any shortfall in the achievement of objectives, appropriate actions are to be taken for improvement in consultation with the management.

The auditor commended the efforts, noting, "Your commitment to embedding the process approach is evident. This is a significant improvement."

## Reflection

That evening, Prakash sat with GM Arun Verma to reflect on the journey. Arun admitted, "I used to think formal processes were just paperwork. Now I see how they drive results and ensure compliance. This has been a wake-up call for us all."

The experience reinforced the importance of aligning operational practices with international standards. For Prakash and his team, it was not just about maintaining certification; it was about building a resilient and effective organization.

Chapter 04

# Audit Nonconformities in Management Processes—Planning

## CS 17 [AS9100D Clause 6.1] Risky Business: Turning Audits into Opportunities

In a bustling Indian ASD company, Management Representative Sanjay Mehta faced his first surveillance audit since joining the organization. The audit uncovered a significant NC under criteria 6.1 of AS9100D/EN 9100:2018:

*Actions to address and manage organizational risks and opportunities are not evident. The risk register identifies only the process risks.*

The auditor recorded appropriate evidence below the NC.

This revelation left Sanjay in a challenging position. The QMS documentation prepared by his team was designed primarily to impress customers and auditors, rather than to address practical issues. Sanjay, who joined the company five years after its establishment, had been primarily focused on ensuring the smooth production of ASD parts and assemblies. The broader aspects of risk and opportunity management had taken a backseat.

### Immediate Containment Action

Realizing the urgency, Sanjay called for an emergency meeting with his team and key department heads. Together, they implemented the following immediate containment actions:

- **Interim Risk Register Update:** The team expanded the existing risk register to include potential organizational risks, such as supply chain disruptions and customer satisfaction issues.
- **Awareness Communication:** A memo was circulated to all managers, emphasizing the importance of identifying and addressing risks and opportunities at an organizational level.
- **Auditor Engagement:** Sanjay assured the auditor that a comprehensive corrective action plan was underway to address the gaps.

## RCA

Sanjay and his team embarked on a detailed RCA. They used the 5 Whys method and identified the following issues:

1. **Limited Understanding of Standards:** The team lacked awareness of AS9100D clauses, especially 6.1 (risks and opportunities), 4.1 (context of the organization), and 4.2 (needs and expectations of interested parties).
2. **Narrow Focus on Process Risks:** Risk management efforts had been confined to process-level issues, ignoring broader organizational risks and opportunities.
3. **Reactive Approach:** The organization's risk management practices were more reactive than proactive, addressing issues as they arose rather than planning ahead.

## Corrective Action Plan

With the root causes identified, Sanjay worked closely with his team to implement a robust corrective action plan:

1. Comprehensive Risk Management Framework:
    - The risk register was revamped to include organizational risks, such as financial challenges, market fluctuations, regulatory changes, and technological advancements.
    - Opportunities, such as expanding into new markets and adopting advanced manufacturing techniques, were also identified.
2. Training and Awareness Programs:
    - Workshops were conducted to educate senior employees about AS9100D clauses 6.1, 4.1, and 4.2.
    - Process owners were trained to align their risk management practices with organizational objectives.
3. Stakeholder Engagement:
    - Input was sought from customers, suppliers, and other stakeholders to understand their expectations and potential risks to the organization.
4. Regular Risk Reviews:
    - A schedule was established for periodic reviews of the risk register to ensure it remained relevant and comprehensive.
    - Risk management became a standing agenda item in management review meetings.

- Risk and opportunities related to the external and internal issues (clause 4.1), and needs and expectations of interested parties (clause 4.2) were extracted from the relevant registers and included in the organizational risk management register kept in the system.

### The Turnaround

Three months later, during the follow-up audit, the organization presented a transformed approach to risk and opportunity management. The updated risk register showcased a balanced view of organizational and process risks, along with actionable plans to address them. Employees demonstrated a clear understanding of their role in identifying and managing risks.

The auditor commended the efforts, stating, "This is a significant improvement. Your proactive approach to risk management aligns well with the intent of AS9100D."

### Reflection

That evening, as Sanjay reviewed the day's success with his team, he remarked, "This experience has taught us the true value of the process approach. Addressing risks and opportunities is not just about compliance—it is about ensuring our organization's long-term resilience and about issues facing the organization."

The journey had been a learning curve for everyone involved, transforming the organization's mindset from reactive to proactive and laying a solid foundation for future success.

## CS 18 [AS9100D Clause 6.2] From Subjective to Smart: Turning Planning into Precision

In the sprawling corridors of an Indian ASD industry organization, the air was tense. The first surveillance audit was underway, and the production planning department was under scrutiny. The process owner, Rajiv Mehta, and his team had spent weeks preparing for this day. Their desks were piled with neatly organized files, flowcharts, and reports, each documenting the processes they had meticulously refined.

When the lead auditor, a composed but sharp-eyed professional named Arjun Rao, entered the department, Rajiv welcomed him with a firm handshake and a confident smile. However, the confidence began to waver as Arjun dived into the details of their operations. "Let's discuss your adherence to clause 6.2 of the AS9100D standard," Arjun began, pulling out his checklist. "Can you show me evidence of your quality objectives for the production planning process?"

Rajiv nodded, motioning to a document. "Here's the process documentation we've prepared, and we've been measuring results based on it."

Arjun examined the document, flipping through its pages. His brow furrowed. "I see your measurements, but they appear to be subjective. There's no clear identification of quality objectives, no periodic measurement or reporting to management. And what actions are taken when targets aren't met?"

Rajiv's stomach sank. He had anticipated questions, but not this level of precision. "We ensure all parts pass inspection and acceptance. Is that not sufficient evidence?"

Arjun shook his head gently.

> *Inspection and acceptance address product quality, not the effectiveness of your planning process. Clause 6.2 requires setting measurable objectives, regularly evaluating performance, and taking corrective actions when objectives are not met.*

The auditor gave the NC in the NCR format. The auditor recorded appropriate evidence below the NC.

The words hit Rajiv hard. The team had focused on process execution but overlooked establishing a structured approach to measuring and improving planning effectiveness. The NC was clear.

As soon as the audit concluded for the day, Rajiv gathered his team in the conference room. "We need to address this NC immediately," he began, his voice steady but urgent. "It is not just about passing the audit; it is about aligning with the standards and strengthening our processes."

The team delved into clause 6.2 of the AS9100D/EN 9100:2018 standard. They realized their gap in understanding—while they had documented processes and subjective assessments, they had not identified objective quality goals, nor had they developed a system to track progress against these goals.

Their first step was containment. Rajiv drafted a temporary framework to identify and monitor quality objectives for the planning process. These included metrics such as on-time completion of production schedules, accuracy of demand forecasts, and adherence to resource allocation plans. He also scheduled weekly meetings to review performance and discuss areas of improvement.

Simultaneously, Rajiv led an RCA to understand how the oversight occurred. They identified a lack of awareness of clause 6.2 among the team and inadequate training in quality management principles.

To address the root cause, Rajiv coordinated training sessions for his team to improve their understanding of AS standards. He also developed a permanent corrective action plan, which included:

1. **Defining Measurable Objectives:** Establishing clear, quantifiable goals for production planning, such as reducing schedule deviations by 10% within six months.
2. **Regular Reviews:** Implement a monthly review cycle to evaluate performance against objectives and report findings to management.
3. **Action Plans for Deviations:** Creating protocols to identify causes of underperformance and take timely corrective actions.

When the auditors returned for verification of the effectiveness of the corrective actions, they were greeted with a transformed department. Arjun reviewed the updated documentation and observed the newly implemented processes in action. Rajiv walked him through the measurable objectives, performance metrics, and evidence of periodic reviews.

"I must say," Arjun remarked, "this is a commendable turnaround. Not only have you addressed the NC, but you've also laid a foundation for continual improvement. Well done."

As the team breathed a collective sigh of relief, Rajiv felt a newfound sense of purpose. The audit had been a wake-up call, a reminder that excellence was not just in meeting standards but in embracing them to drive meaningful progress.

## CS 19 [AS9100D Clause 6.2] Bridging the Gap Between Goals and Actions

Richard Mason, the Quality Management Representative at AeroTechnica, prided himself on his decades of experience with ISO 9001. Known for his meticulous approach, Richard had shepherded the company through countless audits, always ensuring compliance with quality management standards.

But times were changing. AeroTechnica was venturing into the aerospace sector, and with it came the need to achieve AS9100D certification. While Richard attended the transition meetings and read through some materials on AS9100D, he leaned heavily on his ISO 9001 expertise, assuming it would carry over seamlessly.

The internal audit day arrived. The lead auditor, Khadgi Mehta, had extensive experience with aerospace standards and was known for her thoroughness. As the team worked through the audit checklist, everything seemed to progress smoothly—until they reached clause 6.2 quality objectives and planning to achieve them.

Khadgi leaned back in her chair, tapping her pen on the desk. "Richard, let's review the company's quality objectives. Can you show me how you're setting, measuring, and acting upon them?"

Richard confidently pulled out a binder labelled "Quality Goals." It contained pages of general objectives, such as "Improve customer satisfaction" and "Enhance production efficiency."

Khadgi scanned the documents. "These are well intentioned, but where are the measurable targets? For example, for 'Improve customer satisfaction,' what's your baseline score, and what's the target? Also, how do you monitor and act on unmet goals?"

Richard frowned. "We review these during management meetings and discuss progress, but we do not always assign numerical targets. Our assessments are more qualitative. If something feels off, we address it as needed."

Khadgi nodded but made a note in her report. As the audit continued, it became evident that:

- Objectives lacked **specific, measurable targets** (e.g., "Increase OTD by 10% within six months").
- There was no formal process for tracking **progress against these objectives** or verifying corrective actions when targets were not met.
- Reviews of objectives were often subjective and undocumented.

When Khadgi presented her findings, she identified this gap as a significant NC:

*Clause 6.2 B of AS9100D requires a clear connection between objectives, targets, and actions. Without measurable targets, there is no way to evaluate success or take meaningful corrective actions.*

Richard felt a pang of frustration. "This wasn't an issue in ISO 9001. Why is it so critical now?"

Khadgi explained, "In aerospace, the stakes are much higher. Clear, measurable objectives help ensure reliability, safety, and quality. Subjective assessments won't suffice because they can lead to missed risks and inefficiencies. This is about accountability and continual improvement."

Realizing the seriousness of the gap, Richard set a meeting with his team to discuss a plan of action. Together, they agreed on the following steps:

### Follow-Up Actions

1. **Define Measurable Objectives:** Revise all quality objectives to include clear, quantifiable targets. For example, "Achieve 95% OTD within the next quarter" or "Reduce scrap rate by 5% by year end."
2. **Develop Monitoring Processes:** Create a system to track progress, including dashboards or reports updated monthly. Each objective would have a designated owner responsible for monitoring and reporting.
3. **Establish Action Plans for Missed Targets:** Develop a formal process for corrective actions when objectives are not met. This included RCA, action implementation, and verification of results.
4. **Conduct Training:** Organize workshops to educate the team on AS9100D requirements for quality objectives, emphasizing the importance of data-driven decision-making.
5. **Update Management Review Processes:** Integrate objective reviews into regular management meetings, ensuring that progress, gaps, and corrective actions are discussed and documented.

### Outcome

Weeks later, during the external audit, the CB reviewed the updated process. AeroTechnica's revised objectives and robust tracking impressed the auditors, who noted the swift and effective resolution of the NC.

For Richard, the experience was a humbling but valuable lesson. While his ISO 9001 expertise had been a strong foundation, embracing the nuances of AS9100D required a mindset shift. The process not only strengthened AeroTechnica's QMS but also reignited Richard's passion for learning and adapting to new challenges.

## CS 20 [AS9100D Clause 6.3] Change Managed: From Assumptions to Assurance

At Indira Aerospace Systems Pvt. Ltd., the second surveillance audit loomed large. The planning department, led by Arvind Rao, was in the spotlight. Arvind and his team had spent weeks preparing documentation for the audit, confident they had addressed every detail of the production process.

But the confidence wavered when the lead auditor, Sneha Iyer, stepped into the department. Sneha was known for her sharp focus on detail, and this time, her attention was directed at clause 6.3 of AS9100D/EN 9100:2018—change management. All kinds of changes, internal document changes, process changes, customer drawing, order changes, etc.

"Let's review your approach to managing changes in the production process," Sneha began, scanning through their documents. "I understand there was a tool change implemented last month. Can you show evidence that the change was reviewed for effectiveness, ensured no operational disruptions, and didn't affect customer objectives?"

Arvind gestured to a file. "We've documented the approval for the tool change and the process updates. The change was implemented smoothly."

Sneha reviewed the file carefully. Her expression turned serious. "You've documented the change, but there's no evidence of an impact analysis. How did you verify that the tool change didn't disrupt operations or affect your ability to meet objectives? Did you assess whether customer expectations were impacted?"

Arvind faltered. "We didn't encounter any major issues during implementation, so we assumed it was effective."

Sneha nodded, jotting down notes. "Clause 6.3 requires more than an assumption. You must demonstrate that changes are evaluated for risks, monitored for effectiveness, and do not adversely affect operations or customer satisfaction. That evidence is missing here."

*The statement of NC was clear: the planning department lacked a structured approach to evaluating and validating process changes. The auditor recorded appropriate evidence below the NC.*

After the audit, Arvind convened an emergency meeting with his team. "We've overlooked a critical aspect of change management," he admitted. "It is not enough to approve changes—we need to evaluate their impact and document the results."

The team reviewed clause 6.3 in detail and identified the gaps in their process. While they had a system for approving changes, there was no structured method to assess their impact on operations, objectives, and customers.

Their first step was to implement containment actions. Arvind and his team retrospectively analyzed the tool change from the previous month. They reviewed production data to confirm that the change had not disrupted operations or affected quality metrics. They also reached out to key customers to confirm there were no negative impacts on deliveries or product performance.

Next, they conducted an RCA. The issue stemmed from inadequate awareness of the requirements under clause 6.3 and a lack of training on change evaluation practices.

To address the root cause, Arvind developed a corrective action plan:

1. **Standardizing Change Management:** The team created a structured process for managing changes, including mandatory impact assessments before implementation.
2. **Impact Evaluation Metrics:** They defined key metrics to evaluate changes, such as production efficiency, defect rates, and OTD.
3. **Customer Impact Assessment:** A protocol was established to ensure any changes were communicated to customers, with follow-ups to confirm satisfaction.
4. **Training and Awareness:** Arvind arranged workshops for his team to build awareness of clause 6.3 and the importance of evidence-based change management.

When Sneha returned for the follow-up audit, she was greeted with a transformed approach. Arvind presented the new change management process, along with evidence from the retrospective analysis of the previous tool change.

Sneha reviewed the documentation and spoke with team members to verify their understanding of the new process. "This is an impressive improvement," she said, her tone warm. "You've not only addressed the NC but also built a robust system to prevent similar issues in the future."

Arvind felt a surge of pride as his team exchanged relieved smiles. The audit had been a valuable lesson in the importance of structured, evidence-based change management. At Indira Aerospace Systems Pvt. Ltd., the planning department was no longer just managing changes—they were mastering them.

Chapter 05

# Audit Nonconformities in Support Processes

## CS 21 [AS9100D Clause 7.1.3] From Maintenance to Mastery: The Power of Metrics

In the bustling corridors of M/s Tejas Aeronautics Pvt. Ltd., an esteemed organization within the Indian ASD industry, the recertification audit was underway. The audit team meticulously examined every department, seeking alignment with the stringent requirements of AS9100D/EN 9100:2018. The spotlight eventually fell on the maintenance department, helmed by the experienced yet unsuspecting manager and his diligent team.

The maintenance department prides itself on its well-organized processes. Maintenance schedules were carefully drafted, outlining the upkeep of machinery, computers, and servers. A breakdown register was diligently maintained, and operating and maintenance manuals were readily available for reference. Software and antivirus programs were periodically purchased and updated, ensuring the digital infrastructure remained robust. Despite this seemingly structured approach, the audit revealed a critical gap.

### The Unseen NC

During the audit, the team examined clause 7.1.3 of the AS9100D/EN 9100:2018 standard, focusing on infrastructure maintenance. The auditors probed the department's practices, asking pointed questions about the visibility and measurability of maintenance activities. It was here that a glaring NC emerged:

> *"Infrastructure maintenance activities are seldom visible/recorded. There are no set objectives to measure periodically." The auditor recorded appropriate evidence below the NC.*

The manager and his team, though confident in their processes, had overlooked a critical requirement. While they had a solid operational framework, they had not set specific objectives to measure the effectiveness of their maintenance activities. This lack of measurable goals and visibility meant that the maintenance process—an integral part of ensuring operational continuity—was not aligned with the standard's expectations.

### Addressing the Process Challenge

The audit findings left the manager and his team reflecting deeply on their approach. The statement of NC prompted them to revisit their understanding of the AS standard clause. They realized that awareness of the requirement had been insufficient, leading to a gap in their system. Determined to address the issue, they embarked on a structured problem-solving journey.

The team began by conducting a thorough RCA. They identified two key issues:

1. **Lack of Awareness:** The team had not fully comprehended the clause's emphasis on measurable objectives and visibility.
2. **Absence of a Measurement Framework:** While maintenance activities were performed, there were no defined metrics to evaluate their effectiveness.

With these insights, the manager and his team devised a corrective action plan. They:

- **Improved Awareness:** Conducted training sessions to enhance the understanding of AS9100D requirements, particularly focusing on clause 7.1.3.
- **Set Measurable Objectives:** Established clear objectives for maintenance activities, such as reducing downtime by a specific percentage and tracking adherence to maintenance schedules.
- **Enhanced Visibility:** Implemented a digital logging system to record all maintenance activities, ensuring transparency and traceability.
- **Periodic Reviews:** Scheduled regular reviews to evaluate maintenance performance against the set objectives, fostering a culture of continual improvement.

### The Path to Compliance

The implementation of these corrective actions marked a significant turning point for the maintenance department. The next audit cycle revealed a transformed approach—one that not only complied with the AS9100D requirements but also enhanced the department's overall efficiency. The newly established objectives provided clear benchmarks, while the digital logging system ensured every activity was documented and accessible.

The manager, reflecting on the journey, acknowledged the value of the audit's findings. "What initially seemed like a setback," he remarked, "was an opportunity to align our processes with global standards and strengthen our operations."

Through their proactive approach, the maintenance department not only resolved the NC but also set an example for other departments within M/s Tejas Aeronautics Pvt. Ltd., showcasing the power of addressing challenges with determination and a commitment to excellence.

## CS 22 [AS9100D Clause 7.1.4] Cockroach in Space? Not on My Watch!

It was just another Tuesday morning at *AstroBite Foods Pvt. Ltd.*, a proud supplier of food for space missions. Inside the facility, the hum of machinery mixed with the faint buzz of fluorescent lights as employees scurried around, ensuring everything sparkled for the much-dreaded recertification audit.

Ravi, the Production Manager, wore his freshly laundered lab coat like armor. He had overseen the shop floor's meticulous preparation, ensuring the air conditioning unit hummed efficiently, and every surface was cleaner than a five-star restaurant kitchen. The day was a symphony of controlled chaos, and Ravi was its maestro, conducting with precision.

Enter Krishnan, the auditor, with a face as stern as his clipboard. His gaze swept over the shop floor, scrutinizing every inch of the operation. "Shall we begin?" he asked, his tone suggesting that failure was not an option.

Ravi nodded enthusiastically and led Krishnan to the food processing area. Rows of hermetically sealed food pouches gleamed under the lights. Nutrient-packed curries, parathas, and laddoos—all designed for zero-gravity consumption—sat ready to be launched into the cosmos.

For a while, things went smoothly. Krishnan nodded approvingly as Ravi presented temperature logs, humidity records, and airflow reports. But then, it happened.

As Krishnan inspected the underside of a production table, he froze. His pen hovered mid-air. Ravi followed his gaze and saw it—a cockroach, bold as brass, skittering across the floor.

"Ah…" Ravi began, already sweating under Krishnan's piercing stare.

"Do you see that?" Krishnan asked, pointing to the offending insect.

"Yes, yes," Ravi stammered, his mind racing. "It is, uh, part of nature's ecosystem, sir. You know, balance and all that."

"Balance?" Krishnan raised an eyebrow, his clipboard poised for a devastating note.

Before Ravi could respond, the operator at the production line chimed in with surprising nonchalance. "Oh, they come through the drains from the wastewater plant. It is normal."

Krishnan's pen made contact with the clipboard. "Normal?"

"Of course, sir," the operator continued cheerfully. "We've got rodenticides down, and they work 90% of the time. Management even evaluates it continually. The room is dust-proof, and we keep excellent temperature and humidity control."

Krishnan's expression darkened. "So, you're telling me that in a *hygienic process room*, with food destined for *space*, cockroaches are just… accepted?"

Ravi jumped in, desperate to salvage the situation. "We maintain records for everything, sir! And look, the environment is well under control."

Krishnan closed his eyes for a moment, perhaps praying for patience. "AS9100D, clause 7.1.4, clearly states the need for a hygienic environment. Cockroaches do not belong here. This is a major NC."

The words hit Ravi like a meteor. The rest of the audit passed in a haze of polite but firm reprimands.

*Statement of NC: "Determination, provision, and maintenance of the hygienic environment necessary for manufacturing and packaging food is not always likely to achieve quality as cockroaches were seen moving in the shop." (Refer clause 7.1.4 of AS9100D.)*

The auditor had recorded appropriate evidence below the NC.

### The Aftermath

A week later, Ravi and his team sat in the conference room, poring over the NCR. The root cause was clear: insufficient sealing of the drains allowed pests and insects to infiltrate into the facility. Corrective actions were brainstormed and quickly implemented.

The team installed high-grade drain traps, increased pest control measures, and intensified employee training. Posters now adorned the walls, reminding everyone of the unforgiving standards of space food preparation.

When Krishnan returned for a follow-up audit, he found the facility transformed. Not a cockroach in sight.

"Good work," he said gruffly, signing off the report.

Ravi exhaled in relief. He had learned the hard way that in aerospace-grade food production, even the smallest intruder could bring the mightiest operation to its knees.

As Ravi walked back to the shop floor, he made a mental note: space might be the final frontier, but cleanliness would always be the first.

## CS 23 [AS9100D Clause 7.1.5.2] The Landing Gear Lapse: A Calibration Mishap That Took Flight

It was an overcast morning in the bustling aerospace facility when the external auditor Gowda stepped into the metrology lab. The walls, lined with precision instruments and gauges, buzzed with activity as technicians moved between stations. Today's agenda: a routine audit to ensure compliance with AS9100D standards.

After observing the processes for a while, Gowda turned to the metrology technician, Khadgi.

**Auditor:** "Okay, Khadgi, I see how the calibration process flows. Can you tell me how you monitor when a dimensional instrument requires calibration? What's the frequency of calibration for gauges?"

**Khadgi:** "Certainly, sir. Each user is trained on our calibration procedures. They're responsible for bringing their instruments to the lab before the calibration validity expires. If the instrument can't be moved—like the CMM—a calibration technician is requested to perform the service on-site."

Gowda nodded; his brow furrowed slightly. "And how effective is this system in practice?"

Khadgi assured him, "Very effective. We rarely encounter lapses, as users are diligent about following protocol."

Gowda then handed over a list of ten production instruments. "Can you show me the latest calibration records for these?"

Khadgi quickly pulled up the records in the calibration software and handed him the documents.

**Auditor:** (reviewing) "Hmm. Everything seems in order, except... Wait. This report is here. The CMM, ID 9102-AL, used for final assembly, was being operated with an expired calibration."

Khadgi's face tightened. She immediately opened the monitoring software, double-checking the status. "You're right, sir. The calibration period expired last month. It should not have been in use. It seems the user forgot to request calibration."

Gowda's gaze sharpened. "That's a serious issue. Let's assess the impact. Was this instrument used for any critical measurements?"

Khadgi hesitated, then responded, "Yes, sir. We conducted a process failure mode and effects analysis (PFMEA) study, which identified the landing gear as a critical item requiring CMM measurements. I have a copy of the study if you'd like to review it."

The auditor accepted the report, leafing through it with quiet intensity. After a moment, he looked up. "This is a NC."

**Details of NC:**

- Process/Area/Department: Production
- Classification: Major (Ma) Reason: Coordinate-measuring machine (CMM) measurements made on the landing gear are critical and are likely to affect the landing gear functions.
- Requirement/Clause No.: 7.1.5.2

*Statement of NC: "The organization shall establish, implement, and maintain a process for the recall of monitoring and measuring equipment requiring calibration or verification." This process was not effectively implemented.*

The auditor recorded appropriate evidence below the NC as follows:

1. The CMM (ID 9102-AL) was used with its calibration validity expired.
2. The PFMEA study identified landing gear as a critical item requiring accurate CMM measurement.

Later that afternoon, a team of engineers, managers, and metrology experts gathered in the conference room. The NCR lay open on the table, and tension hung in the air.

The Management Representative, Anil, addressed the group. "We need to investigate the root cause and ensure this doesn't happen again. Khadgi, can you brief us on how this oversight occurred?"

Khadgi nodded. "The primary issue seems to be user reliance on manual reminders. If a user forgets to flag an instrument for calibration, it slips through."

Anil turned to the calibration supervisor. "What's your take, Ravi?"

Ravi replied, "The monitoring software can send automatic alerts, but we haven't fully activated those features. We also lack a robust escalation process to catch such oversights."

The team brainstormed corrective actions:

1. Containment Action:
   - Remove the CMM from service until recalibrated and validated.
   - Perform a risk assessment on all measurements taken during the expired period to determine any impact on product quality.
2. RCA and Follow-Up Actions (Short Term):
   - Activate automated alerts for calibration due dates within the monitoring software.
   - Implement a backup system where the calibration team directly reviews due dates weekly.
   - Train all users on the updated process and emphasize the importance of timely calibration.
3. Permanent Corrective Action (Long Term):
   - Integrate calibration tracking with production planning software to prevent scheduling tasks on overdue instruments.

Two weeks later, the auditor returned for a follow-up. The CMM had been recalibrated, and records showed no significant deviations in measurements taken before recalibration. The automated alert system was active, and users had undergone training.

Gowda reviewed the actions taken and smiled. "Good work. This corrective action plan is thorough. I'll verify its long-term effectiveness during the next audit. For now, I'll close this NC."

The room exhaled collectively as the auditor left, leaving behind a clear reminder: in a world of precision, even a moment's lapse can ripple through the skies.

## CS 24 [AS9100D Clause 7.1.5.2] The Missing Details in the Calibration Register

At *PrecisionTech Pvt. Ltd.*, the quality department prides itself on being the gatekeeper of accuracy. Everything, from microchips to precision-engineered bolts, depended on exacting standards. It was their job to ensure all monitoring and measuring equipment was delivered with absolute precision.

When the auditor arrived for the AS9100D recertification audit, the department was confident. Quality Manager Meera escorted the auditor, Devendran Murthy, through the pristine calibration lab, where rows of gleaming instruments stood as testaments to their rigor.

"This is our calibration area," Meera explained with pride. "We ensure all equipment is periodically calibrated to maintain traceability and accuracy."

Devendran Murthy nodded and requested the calibration register. A junior quality inspector handed over a thick logbook. Meera stood tall, ready to defend their process. But as Devendran Murthy flipped through the pages, his expression shifted from neutral to concerned.

"There seems to be a problem here," he said, looking up.

"Problem?" Meera echoed, a faint unease creeping into her tone.

"Yes. While you've recorded which equipment has been calibrated, there's no mention of the calibration method, frequency, or even the acceptance criteria. Without these details, how can you ensure the accuracy of your process or demonstrate traceability to standards?"

Meera's stomach sank. She flipped through the register, seeing for herself what Devendran Murthy had pointed out. It was true—the essential details were missing.

"Well," Meera began cautiously, "we've always followed a schedule and maintained records of when calibration is performed. Surely that counts for something?"

"It is not enough," Devendran Murthy said firmly. "Clause 7.1.5.2 of AS9100D is explicit about these requirements. Without this information, your process lacks the transparency and reliability expected in a QMS."

The audit report later classified this oversight as an NC. Meera and her team knew they had to act swiftly.

***Statement of NC:** "In the register maintained for monitoring and measuring equipment by the organization, information about the calibration or verification method, frequency, and acceptance criteria is absent."*

## Containment and Corrective Actions

The first step was containment. Meera immediately halted the use of any calibrated equipment until verification procedures were reviewed and brought into compliance. Production schedules were adjusted to minimize disruption.

Next came RCA. The team traced the issue to an outdated understanding of calibration record requirements. While past audits had overlooked the missing details, this knowledge gap had persisted, leaving the system vulnerable.

Corrective actions were developed and implemented:

1. A detailed format for the calibration register was designed to include the calibration or verification method, frequency, and acceptance criteria.
2. A training session was conducted for all quality personnel to ensure they understood the requirements of AS9100D.
3. An internal audit schedule was established to regularly verify compliance with the updated procedures.

## The Follow-Up Audit

When Devendran Murthy returned for the follow-up audit, he was presented with the revamped calibration register. The document now included detailed records for each piece of equipment: the calibration method, frequency, acceptance criteria, and even traceability to national standards.

"This is a significant improvement," Devendran Murthy remarked as he signed off the report.

Meera allowed herself a small smile of relief. It had been a tough lesson, but the department emerged stronger, with systems that could withstand the scrutiny of even the most discerning auditor.

## CS 25 [AS9100D Clause 7.1.5.2] Calibration in the Cloud: The Missing Conditions

At *PreciseWorks Pvt. Ltd.*, the quality department operated like a well-oiled machine. Monitoring and measuring equipment were routinely sent to third-party calibration labs, and each returned unit came with a shiny certificate of calibration. For years, this arrangement had been the backbone of their quality assurance process.

When the AS9100D recertification audit began, the Manager of Quality, Sanjay, was confident. He led the auditor, Desai, through the calibration records, neatly organized and bound in a thick file.

"These are the certificates from our approved calibration labs," Sanjay said proudly. "Each lab is audited annually and listed on our supplier approval register."

Desai nodded, flipping through the records. "I see the certificates. But tell me, how do you ensure that the calibration is carried out under suitable environmental conditions?"

Sanjay blinked. "The labs are accredited to ISO/IEC 17025. We trust their processes."

"Yes," Desai said, "but do you verify their adherence to environmental conditions? Calibration is sensitive to factors like temperature and humidity. If conditions deviate, the results may not be valid."

Sanjay's confidence wavered. "We... haven't explicitly verified that. It is implied in their accreditation."

Desai set the file down. "Clause 7.1.5.2 of AS9100D requires you to confirm that calibration is performed under suitable conditions. Without evidence of this, you cannot guarantee the reliability of your measuring devices."

The realization hit Sanjay like a ton of bricks. The certificates were meaningless if the calibration process lacked environmental integrity.

### Containment and RCA

Sanjay immediately convened his team to assess the impact. They identified critical devices whose calibration might have been compromised and performed additional verifications using in-house standards. Non-critical devices were quarantined until the issue was resolved.

During the RCA, the team discovered the oversight stemmed from a lack of detailed communication with the calibration labs. The purchase orders (POs) sent to the labs focused on turnaround time and cost but omitted specific requirements for environmental conditions.

### Corrective Actions

The team devised a three-pronged corrective action plan:

1. Supplier Communication: All future calibration requests to suppliers would explicitly state the required environmental conditions for calibration.
2. Verification Process: A checklist was introduced to verify calibration reports, ensuring they documented the environmental conditions under which the calibration was performed.

3. Supplier Audit Update: The supplier evaluation process was updated to include on-site audits of calibration labs, focusing on their environmental controls.

To prevent recurrence, training sessions were conducted for procurement and quality personnel to emphasize the importance of environmental conditions in calibration.

### The Follow-Up Audit

When Desai returned for the follow-up audit, she reviewed the updated procedures and evidence. She saw clear documentation for recent calibrations, including records of environmental conditions, and reviewed the revised supplier audit reports.

"This is a significant improvement," she said, signing off the report.

Sanjay breathed a sigh of relief. The oversight had been costly, but it had taught the team a valuable lesson: precision was not just about numbers—it was about ensuring every step of the process adhered to the highest standards.

*Statement of NC: "There was no evidence to confirm that the calibration or verification of monitoring and measuring equipment was carried out under suitable environmental conditions. The organization did not ensure this when they requested the service from approved suppliers (calibration labs)."*

The auditor had recorded appropriate evidence below the NC.

## CS 26 [AS9100D Clause 7.1.6] Core Competency: Patricia's Missing Link

Patricia Turner, the HR Manager at AeroTech Industries, had built her career on the foundation of ISO 9001. With over 25 years of experience, Patricia was confident in her ability to ensure compliance with quality standards, particularly in areas related to employee training and competence. However, when the company decided to pursue AS9100D certification to strengthen its position in the aerospace market, Patricia faced a new challenge.

She skimmed through the AS9100D requirements but assumed most of her established HR practices would align seamlessly. It was not until the internal audit that she discovered just how different AS9100D could be.

### The Audit Begins

The lead auditor, Jason Lee, began the session by reviewing HR's role in ensuring employees were equipped with the necessary skills and knowledge to maintain operational excellence. Jason asked, "Patricia, can you walk me through how AeroTech manages the training of recruits and ensures they have the knowledge needed to perform effectively?"

Patricia smiled, confident in her answer. "We've got a robust system. Every new hire goes through orientation and on-the-job training. We rely heavily on our senior employees to mentor them and pass down expertise."

Jason nodded thoughtfully. "That's great for onboarding. But how does AeroTech ensure that critical organizational knowledge—things like proprietary processes, lessons learned, or best practices—is retained and shared systematically? What happens if a key employee leaves?"

Patricia hesitated. "Well, our senior employees are incredibly reliable. We haven't felt the need to formalize their knowledge. It is all in their heads or shared during informal meetings."

Jason made a note and explained, "AS9100D clause 7.1.6 requires evidence that organizational knowledge is identified, retained, and shared. It is not just about individual expertise—it is about ensuring the company's critical knowledge is preserved for continuity and improvement."

## The Gap Is Identified

As the discussion unfolded, Jason pointed out the risks of relying solely on verbal communication and informal mentoring. Patricia realized there was no system in place to:

- Document critical processes or lessons learned.
- Ensure knowledge is consistently shared across teams.
- Retain institutional knowledge when employees leave or transition to other roles.

Jason documented the NC:

*Clause 7.1.6—there is no evidence of organizational knowledge and whether it is retained and shared.*

## The Response

Patricia, though initially caught off guard, quickly recognized the importance of addressing this gap. She convened a meeting with her HR team and other department heads to outline a plan of action.

## Follow-Up Actions

1. Containment Action: Identify key areas of organizational knowledge that were critical to AeroTech's operations, such as specialized manufacturing techniques, customer-specific requirements, and compliance protocols.
2. RCA: Investigate why organizational knowledge had not been documented or shared systematically. The team identified that the issue stemmed from a lack of awareness about AS9100D requirements and over-reliance on experienced employees' verbal instructions.
3. Corrective Actions:
   - Develop a Knowledge Repository: Create a central digital platform where critical knowledge is documented and stored. This includes work instructions, troubleshooting guides, and process maps.
   - Standardize Knowledge Capture: Implement templates and processes for documenting lessons learned from projects, including successes and failures.
   - Knowledge Sharing Mechanisms: Establish regular sessions where employees present insights, share updates, and discuss best practices, with materials archived for future access.
   - Succession Planning: Develop plans for knowledge transfer during employee transitions or retirements to prevent critical information loss.

4. Training and Awareness:
   - Conduct workshops for HR and department managers to educate them about AS9100D's focus on organizational knowledge and its impact on operations and compliance.
5. Monitoring and Improvement:
   - Include organizational knowledge as a topic in management reviews, with metrics to track progress in capturing and sharing knowledge.

### The Outcome

Three months later, AeroTech had implemented a comprehensive knowledge management system. During the certification audit, Jason revisited the clause and was impressed by the company's progress. The knowledge repository and structured sharing practices not only addressed the NC but also enhanced AeroTech's resilience and efficiency.

Patricia reflected on the experience. "This was a wake-up call for me," she admitted. "I've always valued experience, but this taught me the importance of capturing and preserving it for the future. It is not just compliance—it is about ensuring we're stronger as an organization."

The journey to AS9100D certification not only resolved the NC but also elevated AeroTech's operations. For Patricia, it was a lesson in embracing change and turning gaps into opportunities for growth.

## CS 27 [AS9100D Clause 7.2] Bridging the Competency Gap: Training for Success

AeroParts Manufacturing has long been a trusted supplier of built-to-print components for the aviation and defense industries. The company prided itself on precision and reliability, but to secure more significant contracts, it needed AS9100D certification. This certification, tailored to aerospace and defense requirements, demanded not just technical excellence but robust processes for managing personnel competency and training.

Patricia Reynolds, the HR Manager, had prepared meticulously for the internal audit. She was confident that AeroParts' extensive training programs would meet the stringent requirements. However, she had not fully delved into the nuances of AS9100D, assuming her decades of ISO 9001 experience would suffice.

### The Audit Begins

The lead auditor, James Marshall, started his review in the HR department. After verifying training records for the past year, he turned to Patricia and asked, "How does AeroParts ensure that training is not only delivered but also effective in improving competency?"

Patricia smiled and replied, "The competency of employees is reviewed periodically. We ensure all employees undergo regular training sessions specific to their roles. We maintain detailed records of attendance and the content covered."

James nodded, flipping through the files. "This is thorough documentation of who attended and what was taught. But how do you verify that the training has been effective? For instance, how do you ensure that a machinist who attended training on new CNC software is now proficient in its use?"

Patricia hesitated. "We assume that if they attended the training, they've learned what's necessary. If there's a problem, their supervisor usually informs us."

James raised an eyebrow. "AS9100D clause 7.2 requires verification of training effectiveness. Without evidence showing that training has improved employee performance or competency, it is a gap in compliance." James issued the NC: "Clause 7.2—competency/training for people resources was not verified for effectiveness."

Patricia reviewed the findings, realizing that while AeroParts had a strong focus on training, it lacked mechanisms to evaluate whether the training was achieving its goals.

### The Discussion

Patricia, though initially defensive, listened as James explained. "It is not enough to provide training. In the aerospace and defense industries, where precision is paramount, you need to ensure that training translates to enhanced skills and improved performance. For instance, assessments, supervisor evaluations, or on-the-job performance metrics can help verify training effectiveness."

Patricia nodded thoughtfully. "I see your point. We've been focused on providing training but haven't considered measuring its impact. I'll ensure this gap is addressed immediately."

James replied, "That's good to hear. Once you implement containment, conduct an RCA, and establish corrective actions, I'll verify the compliance. If everything is in order, I'll recommend AeroParts for the AS9100D certificate."

### The Follow-Up Actions

Patricia, eager to resolve the NC, convened a meeting with her team and key department heads. Together, they developed a plan:

1. Containment Action:
   - Identify critical roles where training effectiveness needed immediate evaluation.
   - Conduct quick assessments for these employees through written tests, practical demonstrations, or supervisor feedback.
2. RCA:
   - Investigate why training effectiveness was not being verified.
   - Patricia discovered that the focus had been on meeting attendance and content delivery, with no formal requirement to evaluate outcomes.
3. Corrective Actions:
   - **Develop an Evaluation Framework:** Implement a system to assess training effectiveness through post-training assessments, quizzes, and supervisor evaluations.
   - **Integrate Competency Metrics:** Link training programs to KPIs to measure improvements in job performance.
   - **Feedback Mechanism:** Collect feedback from employees and supervisors on the relevance and effectiveness of training sessions.
   - **Regular Reviews:** Schedule periodic reviews of training outcomes during management meetings.

4. Training and Awareness:
   - Educate supervisors and HR personnel on the importance of verifying training effectiveness under AS9100D requirements.
5. Long-Term Monitoring:
   - Maintain a log of training effectiveness evaluations, tying them to competency records for future audits.

### The Outcome

A month later, James returned to verify compliance. Patricia presented the new evaluation framework, along with initial results from employee assessments and feedback. Supervisors provided examples of improved performance, directly linked to the recent training program.

Impressed, James remarked, "This is a robust response to the NC. Not only have you addressed the gap, but you've also enhanced your overall approach to training and competency management. I'm confident AeroParts is ready for AS9100D certification."

### Conclusion

The experience was a turning point for Patricia. "This audit taught me that compliance is not just about checking boxes; it is about ensuring that our processes truly add value. Training is only as effective as the competency it builds."

With the NC resolved, AeroParts secured its AS9100D certification, opening doors to new opportunities in the aerospace and defense sectors. For Patricia, the journey was a testament to the power of listening, learning, and continual improvement.

## CS 28 [AS9100D Clause 7.3 e, f, g, h] Do Not Ignore Sector-Specific Requirements in QMS Upgrades for ASD

The sun peeked through the blinds of the shop manager's office as Arjun sat at his desk, nervously glancing at the clock. Today was the day of the CB audit, and the stakes could not be higher. The Medium Enterprise in which he worked for had been rushing to upgrade their QMS to AS9100D from their ISO 9001 quality system, hoping to break into the high-stakes ASD market.

Arjun had led the upgrade with relentless energy, pushing his team to update procedures, tweak processes, and compile documentation—but all in a hurry. The looming goal was to achieve certification, and the pressure from the board left little room for thorough reviews. He had skimmed through the sector-specific requirements, trusting that the bulk of their ISO 9001 framework would suffice.

The audit began promptly at 9 a.m. The CB auditor, Meera Jones, had a reputation for meticulousness. She navigated through the paperwork and production floors, speaking with employees, observing workflows, and meticulously cross-checking against AS9100D standards.

By late afternoon, the findings were ready. Meera gathered Arjun and the senior leadership team in the conference room. Her expression was serious but not unkind.

## The Audit Findings

Meera cleared her throat and began: "I appreciate the effort your team has put into upgrading the QMS, but there are significant gaps that must be addressed before certification." She slid a document across the table, highlighting the NCR.

The organization had failed to ensure that personnel were adequately aware of key aspects of the QMS, as required by clause 7.3 of AS9100D:

- Relevant documented information and updates.
- Their contribution to product or service conformity.
- Their role in product safety.
- The importance of ethical behavior in the ASD sector.

"During employee interviews," Meera continued, "it became evident that most workers didn't fully understand how their actions impacted safety or quality in the aerospace supply chain. That's a critical oversight."

Arjun's heart sank. The initial internal audits had not flagged this, and the team had assumed documentation updates were enough. But documentation without workforce awareness was like a plane without wings—useless.

## The Path to Recovery

Meera advised immediate containment action. "Pause any production linked to ASD contracts. Address the awareness gap through urgent training and conduct a why-why analysis to find the root cause." She scheduled a follow-up audit in 30 days.

Arjun and his team jumped into action. They conducted the why-why analysis:

1. **Why Did Employees Not Understand QMS Requirements?** Because training sessions were generic and not specific to ASD requirements.
2. **Why Were Training Sessions Generic?** Because the training materials had not been updated to reflect AS9100D's sector-specific requirements.
3. **Why Were Materials Not Updated?** Because the team prioritized document compliance over practical employee understanding.

With the root cause identified, Arjun initiated corrective actions:

- **Training Revamp:** Developed tailored training modules on product safety, ethical behavior, and sector-specific compliance.
- **Documentation Updates:** Created simplified guides for each clause, ensuring every team member could grasp their role in quality and safety.
- **Communication Channels:** Launched regular QMS awareness bulletins and introduced toolbox talks to reinforce learning.

The improvements were swiftly implemented, and employees began to internalize the importance of their roles.

### The Final Verdict

When Meera returned for the follow-up audit, she was impressed. She interviewed workers who confidently explained their contribution to product safety and pointed to clearly updated documentation. The corrective actions were not just paper exercises but had genuinely transformed the organization's culture.

After reviewing the evidence, Meera smiled. "I'm pleased to report that your corrective actions are effective. You've addressed the NCR, and your organization is on track for AS9100D certification with the scope change."

The room erupted in quiet cheers, and Arjun felt a weight lift from his chest. Meera's final words lingered with him:

"This journey has made your organization stronger. The awareness and alignment with ASD standards will not only help you gain certification but also build trust with your customers. Quality is not just about ticking boxes—it is about embedding safety, ethics, and excellence into every layer of your business."

The audit had been a wake-up call, but in the end, it forged a more resilient and knowledgeable team. Arjun left the office that evening, hopeful and determined, knowing that their company was truly ready to soar into the skies of the aerospace industry.

## CS 29 [AS9100D Clause 7.4] From "Not My Job" to "We Got This": A QMS Wake-Up Call

The air was abuzz with excitement in a small Japanese machining industry, known for its precision in crafting turbine blades. The management had recently announced plans to expand into the aerospace sector, a lucrative opportunity that promised to elevate the company's status. However, the journey to aerospace manufacturing required compliance with the stringent AS9100D standard, a step up from their existing ISO 9001 framework. The management assumed the transition would be seamless, but reality hit hard during the certification audit.

The auditor, a seasoned professional with a keen eye, began the session by meeting the management representative. Initial discussions revolved around the establishment of processes and the QMS. However, as the audit delved deeper, a glaring gap became evident. The auditor noticed that shop-floor personnel were unaware of the aerospace-specific requirements. To them, the changes were merely an extension of the existing ISO 9001 system, intended to support business expansion. They were unaware that the new standards prioritized safety, reliability, and maintainability.

When questioned, several workers shrugged off the implementation of AS9100D as "management's job." They admitted to not receiving any formal training on the new requirements. Terms like "product safety," "prevention of counterfeit parts," and "QMS documentation" were foreign to them. The casual approach stemmed from a lack of awareness, which the auditor identified as a critical gap. The auditor also observed that the organization's policy, objectives, and documentation were not effectively communicated, leading to complacency among the staff.

After thorough discussions with the management representative and the quality team, the auditor issued a **NC statement** citing **criteria 7.4 of AS9100D/EN 9100:2018**:

---

*Awareness of all about QMS, personnel and product safety, prevention of counterfeit parts, policy/objectives, and QMS documentation is not effective.*

---

The auditor recorded appropriate evidence below the NC.

This revelation was a wake-up call for the management and quality team. The NC set off a series of urgent follow-up actions.

## Containment Actions

The quality department acted swiftly to contain the issue. Meetings were held with all employees to explain the gravity of the situation. Supervisors were instructed to monitor operations closely to ensure compliance with existing processes, while the gaps were addressed. Interim posters and memos were placed around the facility, summarizing the critical requirements of AS9100D.

## RCA

A cross-functional team conducted a detailed analysis to identify the root cause. They found:

1. **Insufficient Training:** No formal sessions were held to educate employees about the stricter requirements of AS9100D.
2. **Poor Communication:** Management failed to effectively communicate the purpose and significance of the transition to aerospace standards.
3. **Over-Reliance on Management:** The workforce perceived QMS implementation as solely a management responsibility.

## Corrective Actions

Based on the analysis, the following corrective actions were implemented:

- **Comprehensive Training Program:** A robust training module was developed, focusing on aerospace-specific requirements, including product safety, counterfeit prevention, and QMS documentation. All employees, from operators to supervisors, underwent mandatory sessions.
- **Enhanced Communication:** Regular team huddles were introduced to discuss QMS objectives and the importance of adhering to AS9100D. The management also used visual aids, such as posters and videos, to reinforce key messages.
- **Employee Engagement Initiatives:** Workers were involved in process reviews and encouraged to provide feedback on QMS implementation, fostering a sense of ownership and accountability.
- **Comprehensive checklist:** Prepared to prevent the entry of counterfeit parts into the production line.

## Conclusion

By addressing the gaps identified in the audit, the company not only rectified the NC but also fostered a culture of awareness and responsibility. The incident served as a reminder that transitioning to higher standards is not just about upgrading systems—it is about ensuring that every individual in the organization understands and contributes to the collective goal of safety, reliability, and excellence.

## CS 30 [AS9100D Clause 7.5.3.1] The Quest for Perfection: An AS9100D Audit Journey

The factory floor buzzed with activity as Jake, the Management Representative, strode purposefully into the conference room. The recent AS9100D audit had revealed cracks in their processes, and Jake knew the stakes were high. The medium-scale Indian industry was gearing up to supply aerospace components to defense customers, and their QMS had to meet the stringent requirements of the standard.

At the head of the table, the MD waited, his expression tense. The lead auditor began the closing meeting with a somber tone.

"We have identified a significant NC under clause 7.5.3.1," he said, glancing at his notes. "Updating external documents, controlling revisions, and ensuring the latest versions are available at workstations are not effective. My team found multiple examples across processes. Had we documented each issue, it would have resulted in several minor NCs. Instead, we consolidated them into one major finding. I have recorded appropriate evidence below the NC."

The room grew still as the examples were laid out: outdated work instructions at assembly lines, mismatched revision numbers on production floor documents, and suppliers working with obsolete specifications.

"This," the lead auditor concluded, "is a systemic issue that demands immediate attention."

### The Immediate Response

The MD turned to "We cannot let this jeopardize our certification or our reputation. What is the plan?"

Jake, ever prepared, spoke confidently. "First, we will ensure containment by halting all processes impacted by outdated documents. Then, we will revise our documentation control procedures, retrain employees, and deploy a robust document control system. I will personally educate all employees about the importance of AS9100D requirements."

The action plan was swiftly set into motion.

1. **Containment:** All outdated documents were identified and withdrawn from circulation. Temporary workarounds were put in place to avoid disruption.
2. **RCA:** The root cause was traced to a lack of centralized oversight in document control and inadequate training for process owners.
3. **Corrective Actions:**
    - A dedicated document control team was established.
    - A digital document management system with version control was implemented.
    - Employees were trained on identifying and using current versions of documents.

The findings and corrective actions were meticulously documented in the NCR and shared with the auditor.

### The First Follow-Up Audit

One month later, the lead auditor returned to verify the corrective actions. Jake presented a comprehensive report and led the auditor through the factory floor to showcase the improvements.

However, the auditor's sharp eye spotted an outdated work instruction in a remote workstation. "This," he said, "indicates the issue persists. I cannot close this NC. It escalates to a major NC."

The setback was a blow. The MD called an urgent meeting with Jake and the process owners. "We cannot afford half-measures," he said firmly.

### The Rigorous Overhaul

Determined to resolve the issue, Jake intensified the efforts:

- **Comprehensive Audit:** Internal auditors reviewed every document across all locations.
- **Employee Engagement:** Workshops and one-on-one sessions reinforced the importance of document control.
- **Accountability:** Process owners were made directly responsible for document compliance in their areas.
- **Validation:** Weekly audits ensured no outdated documents remained in use.

The NCR was updated with evidence of the revised actions and shared with the lead auditor.

### Certification Achieved

During the next follow-up audit, the lead auditor found no instances of outdated documents. "Your corrective actions have been effective," he said, "and I recommend initial certification. However, I will verify the sustained effectiveness during the next annual audit."

The MD, Jake, and the process owners sighed in relief. The journey had been arduous, but the lessons learned had strengthened the organization's QMS.

Jake addressed the employees in a final meeting. "This certification is just the beginning. Maintaining it demands daily commitment. Together, we can ensure our QMS remains a benchmark in the aerospace industry."

The factory floor erupted in applause, ready to take on the challenges of their new, certified status.

## CS 31 [AS9100D Clause 7.5.3.2] Correcting Records: Prove Your Integrity

The audit had been intense, and as Management Representative Andrew entered the conference room, the atmosphere was charged with anticipation. The lead auditor sat at the head of the table, flanked by his team, ready to present the audit findings to the organization's leadership.

"We have identified a NC under clause 7.5.3.2 of AS9100D," began the lead auditor, his tone firm yet constructive. "Retained records are not adequately controlled. Overwriting, masking, and unauthorized corrections were observed across various processes. In some cases, electronic records were also found to be vulnerable to manipulation."

The auditor recorded appropriate evidence below the NC.

The examples he presented painted a concerning picture:

- Production logs with visible overwriting.
- Quality inspection records with corrected entries lacking authorization or explanation.
- Electronic records without adequate version control or audit trails.

"This is a systemic issue," the lead auditor continued. "Retention of records is foundational to a robust QMS. If I were to log every instance, we would see multiple NCs across processes."

The MD turned to the Management Representative, his expression a mix of disappointment and urgency. "We need a clear action plan; this cannot continue."

### The Response

Andrew rose to the challenge, addressing the room with conviction. "Our goal is to build a near-perfect QMS that meets the stringent requirements of AS9100D and positions us as a reliable supplier of defense aerospace components. This NC is an opportunity to strengthen our processes."

The team quickly moved into action:

1. Containment Actions:
   - Halted processes using records with unauthorized corrections.
   - Designated temporary oversight for critical records to prevent further issues.
2. RCA: Identified the root causes:
   - Lack of training on record-keeping protocols.
   - Absence of a system to authorize and track corrections.
   - Weak controls for electronic records.
3. Corrective Actions:
   - **Standardized Record-Keeping Protocols:** Introduced a documented procedure for corrections, including approval workflows and mandatory annotations for changes.
   - **Training and Awareness:** Conducted training sessions for all employees on the importance of record integrity and the requirements of AS9100D.
   - **Electronic Record Controls:** Implemented software with version control and audit trail capabilities; and restricted access to sensitive records to authorized personnel only.

### Reporting and Verification

Andrew prepared a detailed NCR, including:

- Examples of the identified issues.
- RCA findings.
- Evidence of corrective actions, such as training attendance logs, updated procedures, and screenshots of the new electronic record system.

The auditor returned for a follow-up verification. Andrew walked the team through the changes, presenting updated records, procedures, and system controls.

The lead auditor inspected production logs, quality records, and the electronic database. Satisfied with the improvements, he stated, "Your corrective actions address the NC. I will recommend initial certification. However, the effectiveness of these actions will be verified during the next audit."

### Sustained Improvement

With the NC resolved, Andrew addressed the team in a meeting. "Achieving certification is a milestone, but maintaining it demands vigilance. Every record we keep reflects our commitment to quality and integrity. Together, we will ensure our QMS remains a benchmark in the aerospace industry."

The employees, now trained and empowered, returned to their workstations with renewed purpose, ready to uphold the rigorous standards of AS9100D.

## CS 32 [AS9100D Clause 7.5.3.2] Journey to Compliance: A Management Representative's Determination to Secure AS9100D Certification

In a sprawling ASD organization in India, Rajesh, the Management Representative, stood at the helm of quality compliance. A seasoned mechanical engineer, he had successfully led his team to upgrade their QMS to meet AS9100D requirements. His goal was to maintain certification post-CB audit. However, he found himself grappling with an unforeseen challenge: the transition to controlling electronic documents and records.

### The Audit Day Revelation

The organization had long relied on traditional methods for managing documents and records. But with a growing number of computers and employees eager to access information digitally, the landscape was shifting. Tier 1 and Tier 2 ASD clients required stringent controls for electronic data. Despite warnings, Rajesh hoped to pass the audit with minimal focus on this area, believing the auditors would guide them post-audit.

The audit commenced, and everything seemed to progress as usual. But as Rajesh feared, the auditor zeroed in on the organization's electronic document controls.

"What controls have you established for your electronic data, documents, and records?" the auditor asked, his gaze steady.

Rajesh hesitated. The truth was their electronic controls were rudimentary at best. The audit proceeded, and the inevitable happened—the auditor issued an NCR, stating **the controls were ineffective**. Vide clause No. 7.5.3.2 e,

"When documented information is managed electronically, data protection processes are not clearly defined (e.g., protection from loss, unauthorized changes, unintended alteration, corruption, physical damage)."

### A Plea for Knowledge

Rajesh, humbled but eager to learn, pleaded with the auditor after the audit.

"I'm a mechanical engineer," Rajesh confessed. "I do not have expertise in electronic data control. Can you at least guide me to the right resources or agencies?"

Kunal Nigam, the auditor, and a former project manager at a reputed software design and engineering services firm, agreed to spare a few minutes for Rajesh. He said, "I am happy to share knowledge, but I will give no solution to your issues. Solution is not my cup of tea."

After the audit time, the auditor outlined essential electronic document control measures:

- **Encryption:** Scrambling data to prevent unauthorized access.
- **Setting Strong, Unique Passwords on Documents and Systems:** To prevent unauthorized access while ensuring authorized users can access information securely.
- **Access Controls:** Limiting who can view, edit, or delete documents.
- **Regular Backups:** Safeguarding against data loss, including disaster management and hardware failures, accidental deletions, or cyberattacks.
- **Antivirus and Software Updates:** Protecting systems from threats.
- **Document Management Systems and Cloud Storage:** Organizing and securing documents

and storing documents on secure cloud platforms with built-in security features.

- **Network Controls:** Network controls are security measures implemented within a network to manage and restrict access, monitor traffic, and prevent unauthorized users or devices from accessing sensitive data, acting as a safeguard to protect the network from cyberthreats.
- **Using Firewalls:** IDS, VPNs, and segmentation to monitor and secure traffic.

The auditor clarified he could not prescribe a solution but urged Rajesh to act swiftly, as the deadline for corrective action looms closer with each passing day. The auditor also recommended Rajesh to have a look at the standard ISO/IEC 27001:2022 information security, cybersecurity and privacy protection—information security management systems—requirements.

## The Transformation

Determined not to let the organization falter, Rajesh assembled a task force. They partnered with an external cybersecurity agency to design a foolproof electronic document control system. The team worked tirelessly, implementing encryption, access controls, regular backups, and a state-of-the-art document management system. Firewalls and IDS systems were configured, and staff underwent rigorous training on data security protocols.

RCA identified the organization's over-reliance on manual processes as the key issue. The corrective action plan involved a complete digital transformation, aligning every control with AS9100D standard requirements.

## The Follow-Up Audit

When the auditor returned for the surveillance audit, he meticulously reviewed the new system. He tested encryption protocols, verified access logs, and inspected backup procedures. After thorough verification, the auditor turned to Rajesh with a rare smile.

"I'm impressed," he said. "Not only have you addressed the NCR, but you've also future-proofed your organization's data security. The improvements enhance compliance, reduce cybersecurity risks, and boost customer confidence. You've set a benchmark for AS9100D electronic data control."

Rajesh beamed with pride, knowing their efforts had paid off. The auditor's words were a testament to their resilience and dedication.

## The Organizational Gains

The auditor listed the overall benefits:

- **Enhanced Security:** Reduced risk of data breaches and cyberattacks.
- **Operational Efficiency:** Quicker, more reliable access to documents.
- **Customer Trust:** Strengthened relationships with prime ASD clients.
- **Regulatory Compliance:** Full alignment with AS9100D and industry standards.
- **Future Readiness:** Scalable infrastructure ready to adapt to evolving threats.

Rajesh reflected on the journey. The NCR had been a turning point, propelling the organization toward excellence. And as they celebrated their successful audit outcome, Rajesh silently thanked the auditor for the invaluable guidance—a few minutes that had reshaped their future.

Their story was no longer one of oversight but of transformation, resilience, and a relentless pursuit of quality in the ever-evolving world of aerospace and defense.

Chapter 06

# Audit Nonconformities in Product Realization Processes

## CS 33 [AS9100D Clause 8.1] The Auditor's Eye: Revealing Gaps in Aircraft Lifecycle Management

In a bustling aerospace manufacturing hub in India, the head of production planning and control (PPC) at a large ASD organization sat down for an intense session with an auditor, who was tasked with evaluating the organization's readiness for initial certification to AS9100D. This was a pivotal moment, not only for the company's QMS but also for ensuring that their production processes aligned with the stringent international standards required for manufacturing fighter aircraft for the Indian military.

The audit was comprehensive. The auditor, an experienced and meticulous professional, had studied the detailed D&D documents for the aircraft, scrutinized the production processes, and reviewed the type certification and airworthiness records handed over by the regulatory authority. However, as the audit session progressed, the auditor began to probe deeper into a critical yet often overlooked aspect of the aerospace manufacturing process: lifecycle management of the aircraft.

### The Critical Question

As the PPC head confidently explained the process followed by their department for manufacturing the fighter aircraft, the auditor raised a thought-provoking question: "While determining the product and service requirements from the customer and regulatory authority, what considerations did you account for in the lifecycle management of the aircraft?"

 

The PPC head paused. The question was specific. The auditor was referring to aspects such as:

- Selection and Development of Embedded Software: The role of software in ensuring operational safety and efficiency over the aircraft's lifecycle.
- Electronic Parts/Product Obsolescence: The challenges and strategies for managing components that may become obsolete during the service life of the aircraft.
- Prevention, Detection, and Removal of Foreign Objects: A critical focus to ensure the aircraft's performance and safety are not compromised by foreign object debris (FOD).
- Recycling or Final Disposal of the Aircraft: The ethical and environmentally responsible end-of-life disposal of the aircraft, in line with sustainable practices.

These aspects, the auditor explained, were not just theoretical considerations. They were essential for post-delivery support, especially for a product as sophisticated and high stakes as a fighter aircraft.

The PPC head, however, seemed unprepared for the auditor's pointed questions. After all, their focus had primarily been on the immediate manufacturing process, guided by the design documents and production specifications. The standard's notes on lifecycle management were dismissed as "not applicable" to their specific context. The PPC head assured the auditor that these were matters beyond their scope and were not considered part of the planning process.

### The Auditor's Intervention

The auditor, though understanding the PPC team's stance, was far from convinced.
He explained that while it was true that the notes in AS9100D were not mandatory for the manufacturing phase, they served an important purpose—they called attention to areas requiring consideration during post-delivery support, including obsolescence management, FOD, and final disposal. These factors were not only essential for the aircraft's continued airworthiness but were integral to ensuring long-term sustainability and customer satisfaction.

At this point, the auditor realized that the PPC team had not yet developed a comprehensive plan to address these crucial elements. As a result, he issued a minor NC under clause 8.1 (a), which highlighted a significant gap. The NC pointed out that the PPC team had not considered the lifecycle management aspects during the planning phase, specifically focusing on obsolescence of electronic parts, the management of foreign objects, and the aircraft's end-of-life disposal.

### Containment Action

The PPC head immediately recognized the need to act. In the face of the audit findings, the first step was to implement containment actions. The PPC team swiftly conducted an internal review of the entire planning process. They decided to set up an immediate internal task force comprising cross-functional experts from engineering, production, and after-sales support. The task force's responsibility was to perform a quick assessment of the current situation and identify short-term solutions.

To address the product obsolescence issue, the team began working with the engineering department to ensure that they would implement strategies for managing parts that could potentially become obsolete during the aircraft's service life. They also initiated discussions with suppliers to identify critical components and plan for alternative sourcing or upgrades well in advance.

In terms of foreign object prevention, the team devised immediate measures to tighten controls on the production floor, ensuring no foreign debris could enter the aircraft system during assembly. Additionally, the team coordinated with the quality assurance department to implement more robust inspection processes to detect and remove foreign objects during pre-delivery checks.

To begin addressing the end-of-life disposal, they contacted environmental consultants to assess potential recycling options for the aircraft's materials and identify sustainable disposal strategies for hazardous materials.

### RCA

While the containment actions were swiftly put in place, the real work was to determine the root cause of why these critical lifecycle management elements had been overlooked in the first place. The RCA revealed several factors:

- Misinterpretation of the Standard: The PPC team had misunderstood the notes in the AS9100D standard, dismissing them as irrelevant to the production process, focusing instead on the immediate requirements for manufacturing.
- Lack of Awareness of Post-Delivery Support Needs: There was a general lack of awareness within the PPC team about the importance of long-term support and sustainability considerations, especially in the defense aerospace sector.
- Siloed Departmental Functioning: There was a disconnect between the manufacturing, design, and after-sales support teams, which led to poor communication and a lack of integrated planning for the product lifecycle.

To address these root causes, the PPC department developed a new integrated planning framework. This included regular cross-functional meetings to ensure that all aspects of the product lifecycle—from embedded software to obsolescence management and disposal—were considered during the initial stages of planning.

### Corrective Actions and Auditor's Return

Over the following months, the PPC department worked closely with external agencies to implement the revised procedures for managing lifecycle considerations, including the control of documents, data, and records in electronic media. They ensured that comprehensive records were maintained for every phase of the aircraft's lifecycle, from production to disposal. The records were stored securely in an electronic document management system, ensuring easy access for audits and traceability.

When the auditor returned for his follow-up audit, he found that the PPC team had made significant progress. The new processes were fully integrated into the production planning process, and the corrective actions had been thoroughly documented. The auditor reviewed the evidence and was impressed by the level of detail and the proactive approach the organization had taken to address the NC.

### The Auditor's Compliment and Organizational Gains

The auditor complimented the organization on the effective corrective actions and the improvements in their PLM. Not only did the organization now comply with the requirements of AS9100D, but they had also established a more holistic and sustainable approach to manufacturing their fighter aircraft.

The organizational gains were substantial:

- Enhanced Product Support: The new planning processes ensured that post-delivery support, including obsolescence management and FOD prevention, was integrated into the product lifecycle from the outset.
- Risk Mitigation: By addressing obsolescence and planning for sustainable disposal, the organization mitigated risks associated with long-term product support and maintenance.
- Improved Supplier Relationships: The efforts to manage product obsolescence also fostered better relationships with key suppliers, ensuring continued access to critical components.
- Customer Satisfaction: Ultimately, these changes enhanced the organization's ability to deliver high-quality, reliable products, leading to increased customer satisfaction.

The PPC head, once caught off-guard, now had a sense of pride in how the organization had turned a potential shortcoming into an opportunity for growth. The journey from audit findings to corrective actions not only ensured compliance but also set the organization on a path of continual improvement, strengthening their reputation as a leading aerospace manufacturer.

## CS 34 [AS9100D Clause 8.1.1] The Risk Factor: Can You Manage What You Cannot Measure?

The planning room was tense as the audit of the medium-scale industry neared its conclusion. The planning manager, sweating slightly despite the air conditioning, sat across from the lead auditor. The discussion had been thorough, lasting over two hours, and now the auditor was presenting a critical finding.

"Your process risk identification and management are not effective," the lead auditor began. "While the PFMEA format is being used, there are several gaps in your approach. Mitigation actions lack target dates, and there is no indication of whether the actions are completed. Team responsibilities are not defined, and no date for the next review is planned. Moreover, the maximum acceptable risk is set at 9, which is alarmingly high for aerospace components."

The auditor had recorded appropriate evidence below the NC.

The planning manager nodded grimly, aware of the severity of the issue.

### The Response

The next morning, Amanda Jones (Management Representative) called an urgent meeting with the planning manager and other process owners. "Risk management," Amanda said, addressing the group, "is the backbone of a successful QMS. If we cannot manage risks effectively, we jeopardize not only our certification but also our reputation as a supplier of defense aerospace components."

Amanda outlined the key deficiencies highlighted during the audit and proposed a plan of action:

1. Immediate Containment:
   - All ongoing projects were reviewed to identify high-risk areas with insufficient mitigation plans.
   - Temporary controls were implemented for identified risks to prevent escalation.
2. RCA: Amanda facilitated a workshop with the planning manager and team to uncover the root causes of the NC:
   - Lack of training on the AS9100D risk management requirements.
   - No structured timeline or accountability for mitigation actions.
   - Inadequate tools for tracking risk reviews and mitigation progress.
3. Corrective Actions:
   - **Redesign the PFMEA Format**: Added columns to document mitigation action target dates, responsible team members, completion status, and next review dates.
   - **Training**: Conducted a training program for all relevant employees on effective risk management, focusing on AS9100D requirements.
   - **Risk Acceptance Criteria**: Revised the maximum acceptable risk threshold to a more conservative value based on aerospace industry standards.
   - **System Implementation**: Introduced software for real-time tracking of risks, mitigation actions, and review schedules.

## Reporting and Verification

Amanda meticulously documented the corrective actions in the NCR, including:

- Updated PFMEA formats with sample entries.
- Attendance records and materials from the training program.
- Evidence of revised risk thresholds and the new tracking system.

The auditor reviewed the corrective actions online and found the updates comprehensive and practical. "Your team has made significant strides in addressing this NC," he said during a follow-up call. "I will recommend initial certification. However, the effectiveness of these changes will be evaluated during the next audit."

## Sustaining Robust Risk Management

Following the successful resolution of the NC, Amanda addressed the entire workforce in a company-wide meeting in her role as Management Representative. "Risk management is not just about audits," Amanda emphasized. "It is about protecting our business, our employees, and our customers. Every one of us plays a role in identifying and managing risks from the beginning to the end of production and delivery."

The planning manager implemented regular reviews of the risk management process, ensuring that mitigation actions were tracked, completed, and documented.

## A Stronger Foundation

By the next audit, the organization had transformed its approach to risk management. The lessons learned from the NC laid a stronger foundation for their QMS, positioning the company as a reliable and quality-conscious supplier of aerospace defense components.

## CS 35 [AS9100D Clause 8.1.1] Risking It Right: From NC to Success with a Solid Safety Net

AeroPro Industries, a medium-scale manufacturer located in India, was on the brink of transformation. With aspirations to supply aerospace components to defense customers, the company needed to secure ISO 9001/AS9100D certification, a critical milestone for entering the high-stakes aerospace sector.

The certification audit began with a detailed review of the planning processes. The Head of Planning, Arun Sharma, a seasoned professional with a reputation for strategic thinking, knew this was a defining moment for AeroPro Industries.

### The Audit and a Stark Realization

The audit focused on clause 8.1.1 of the ISO standard, which addresses operational planning and control, including risk management. After an exhaustive two-hour session, the auditor identified a significant NC:

- Potential problems during production and delivery were not fully deliberated.
- Team-based risk evaluations were missing.
- No benchmarks existed for assessing the effectiveness of mitigation actions.
- Residual risks were improperly considered "zero" after mitigation, a clear misunderstanding of risk management principles.

The lead auditor provided candid feedback: "Risk can never be eliminated, only reduced. What's missing here is a systematic approach to identifying, assessing, and mitigating risks. Without this, the process remains vulnerable."

### The Call for Action

Although the feedback was critical, Sharma saw it as an opportunity. Gathering his team, he emphasized the importance of implementing a structured and internationally recognized risk management framework.

"Certification is just one part of the journey," Sharma began. "Our real goal is to build a resilient process that ensures quality and reliability for aerospace customers. Let's turn this NC into a blueprint for excellence."

### Implementing a Robust Risk Management Framework

With urgency and determination, AeroPro Industries adopted the following steps:

1. Aligning with ISO 31000 Standards: The team adopted ISO 31000, the international standard for risk management, to guide their framework. The process involved:
   - **Risk Identification:** Using systematic tools like brainstorming sessions and checklists, they identified risks across procurement, production, quality control, and delivery stages.
   - **Risk Analysis:** Each identified risk was analyzed for its likelihood and potential impact using a risk matrix.
   - **Risk Treatment:** Mitigation strategies were designed, including preventive actions and contingency plans.
2. **Implementing PFMEA:** To address risks specific to manufacturing processes, PFMEA was introduced:
   - Teams analyzed each step of production, identifying potential failure modes and their causes.

- Each failure mode was assigned a risk priority number based on severity, occurrence, and detectability.
- High-priority risks were targeted for immediate mitigation actions.

3. **Solving Potential Problems Through Cross-Functional Teams:** Weekly workshops were held with representatives from production, procurement, and quality. They:
   - Reviewed historical data to identify recurring issues.
   - Brainstormed potential future problems.
   - Established benchmarks for success, ensuring mitigation actions were measurable and effective.
4. **Residual Risk Assessment:** Instead of assuming zero risk post-mitigation, the team calculated residual risks using a quantitative approach. They established thresholds for acceptable risks and monitored them continually.
5. **Continual Monitoring and Evidence:** AeroPro implemented a live dashboard for real-time tracking of risks, mitigation measures, and their effectiveness. Evidence was meticulously documented, including:
   - Updated risk registers aligned with ISO 31000.
   - PFMEA worksheets.
   - Meeting minutes highlighting collaborative problem-solving.
   - Residual risk analyses and thresholds.

## The Follow-Up Audit

Management Representative Sharma documented all actions in the NCR and submitted it to the auditor. During an online verification, the auditor reviewed the evidence, including the risk register, PFMEA records, and action plans.

Impressed by the comprehensive approach, the auditor remarked, "AeroPro Industries has demonstrated a commitment to world-class risk management. I recommend certification. We'll verify the effectiveness during the next audit."

## A New Beginning

With the initial certification secured, AeroPro Industries was now poised to meet the stringent demands of aerospace customers. Sharma addressed his team during the celebratory gathering:

"This is more than a certification. It is a testament to what we can achieve together. With our risk management framework in place, we're not just ready to meet expectations—we're ready to exceed them."

As AeroPro Industries moved forward, its robust risk management framework became a cornerstone of its operations, ensuring quality, reliability, and resilience in an ever-demanding industry.

## CS 36 [AS9100D Clause 8.1.2] Configuring Success: From NC to Excellence

In the bustling industrial town of Pune, India, a medium-scale manufacturing unit, Stellar Aerospace Components Pvt. Ltd., was making its mark in the aerospace sector. The factory floor buzzed with activity, producing high-precision components for various industries. However, the company's ambition to expand into the defense aerospace market faced a critical hurdle—an ineffective configuration management system (CMS).

### The NC

An external audit was underway, with the Planning Manager Arvind Kulkarni as the process owner. After two intense hours of scrutinizing processes, the auditor highlighted a significant NC: "No clear understanding of configuration management (identification and traceability aspects) of individual parts for ASD customers. The process is not effective. Refer to ISO 10007."

This was a pivotal moment for Arvind. The findings could derail Stellar Aerospace's aspirations to enter the defense market. He knew immediate action was required to not only address the NC but also to establish a robust CMS that would assure customers of the company's reliability and compliance.

### The Action Plan

Arvind called an emergency meeting with his team. "We have a clear mission now," he said, his voice firm. "To build a world-class configuration management process that ensures every part we deliver meets the exact requirements of our customers."

He outlined a three-phase action plan:

1. Understanding and Training:
    - Conduct workshops for all stakeholders, from marketing to production, on the principles of configuration management as per ISO 10007.
    - Engage an external consultant specializing in aerospace standards to guide the team.
2. Process Redesign:
    - Develop a configuration management plan (CMP) that includes:
        - Capturing customer requirements through detailed marketing and engineering documentation.
        - Linking customer specifications to part numbers, drawings, and technical data.
        - Implementing software to ensure real-time tracking and traceability.
    - Establish a CCB to review and approve changes at every stage.
3. Implementation and Audit:
    - Roll out the new process across departments.
    - Conduct internal configuration audits to ensure compliance and effectiveness.

### Implementation

The first step was training. Over a week, teams attended sessions led by the consultant, gaining a clear understanding of configuration identification, control, status accounting, and verification.

Next came the process redesign. Marketing teams were equipped with templates to document customer requirements in detail. Engineering teams developed a system to assign unique configuration baselines to each part. A digital configuration management tool was integrated into the enterprise resource planning (ERP) system, ensuring seamless traceability from order placement to delivery.

The CCB met weekly to review change requests, ensuring every modification was documented and approved. The internal audit team conducted mock configuration audits, identifying and addressing gaps.

### The Follow-Up Audit

Six months later, during a follow-up audit, the auditor examined the corrective actions taken. Arvind confidently presented the evidence:

- Training records and participant feedback.
- The newly implemented CMP and its practical application.
- Reports from internal configuration audits.
- Records of configuration control decisions by the CCB.

The auditor verified the actions online, scrutinized the records, and reviewed the traceability of several parts delivered to customers. Satisfied, the auditor remarked, "The process is now robust. Initial certification is recommended."

### The Road Ahead

With a certification in hand, Stellar Aerospace Components began fulfilling orders for defense aerospace customers. The CMS became a cornerstone of their operations, ensuring every component met exact specifications.

A year later, during the surveillance audit, the effectiveness of the corrective actions was verified. The auditor's report was glowing, marking Stellar Aerospace as a trusted partner in the defense sector.

For Arvind Kulkarni, the journey was more than a professional achievement. It was a testament to the power of decisive leadership, teamwork, and a commitment to excellence. Stellar Aerospace's robust CMS was not just a response to an audit finding but a foundation for sustained success in a competitive industry.

## CS 37 [AS9100D Clause 8.1.2] Mission Configured: Turning Audit Findings into Aerospace Triumph

In the bustling industrial town of Pune, India, a medium-scale manufacturing unit, Stellar Aerospace Components Pvt. Ltd., was making its mark in the aerospace sector. The factory floor buzzed with activity, producing high-precision components for various industries. However, the company's ambition to expand into the defense aerospace market faced a critical hurdle—an ineffective CMS.

### The NC

An external audit was underway, with the Planning Manager Arvind Kulkarni as the process owner. After two intense hours of scrutinizing processes, the auditor highlighted a significant NC: "Configuration audit is not done for any part delivered to ASD customers. Refer to ISO 10007." The auditor has recorded appropriate evidence below the NC.

This was a pivotal moment for Arvind. The findings could derail Stellar Aerospace's aspirations to enter the defense market. He knew immediate action was required to not only address the NC but also to establish a robust CMS that would assure customers of the company's reliability and compliance.

### The Action Plan

Arvind called an emergency meeting with his team. "We have a clear mission now," he said, his voice firm. "To build a world-class configuration management process that ensures every part we deliver meets the exact requirements of our customers."

He outlined a three-phase action plan:

1. Understanding and Training:
    - Conduct workshops for all stakeholders, from marketing to production, on the principles of configuration management as per ISO 10007.
    - Engage an external consultant specializing in aerospace standards to guide the team.
2. Process Redesign:
    - Develop a CMP that includes:
        - Capturing customer requirements through detailed marketing and engineering documentation.
        - Linking customer specifications to part numbers, drawings, and technical data.
        - Implementing software to ensure real-time tracking and traceability.
    - Establish a CCB to review and approve changes at every stage.
    - Introduce mandatory configuration audits to verify compliance before delivery.
3. Implementation and Audit:
    - Roll out the new process across departments.
    - Conduct internal configuration audits to ensure compliance and effectiveness.
    - Document findings and corrective actions in the NCR format.

### Implementation

The first step was training. Over a week, teams attended sessions led by the consultant, gaining a clear understanding of configuration identification, control, status accounting, and verification.

Next came the process redesign. Marketing teams were equipped with templates to document customer requirements in detail. Engineering teams developed a system to assign unique configuration baselines to each part. A digital configuration management tool was integrated into the ERP system, ensuring seamless traceability from order placement to delivery.

The CCB met weekly to review change requests, ensuring every modification was documented and approved. Configuration audits were conducted for all parts before delivery, ensuring strict compliance with customer specifications. The internal audit team conducted mock configuration audits, identifying and addressing gaps.

### The Follow-Up Audit

Six months later, during a follow-up audit, the auditor examined the corrective actions taken. Arvind confidently presented the evidence:

- Training records and participant feedback.
- The newly implemented CMP and its practical application.

- Reports from internal configuration audits.
- Records of configuration control decisions by the CCB.
- Evidence of configuration audits performed for parts delivered to ASD customers.

The auditor verified the actions online, scrutinized the records, and reviewed the traceability of several parts delivered to customers. Satisfied, the auditor remarked, "The process is now robust. Initial certification is recommended."

### The Road Ahead

With certification in hand, Stellar Aerospace Components began fulfilling orders for defense aerospace customers. The CMS became a cornerstone of their operations, ensuring every component met exact specifications.

A year later, during the surveillance audit, the effectiveness of the corrective actions was verified. The auditor's report was glowing, marking Stellar Aerospace as a trusted partner in the defense sector.

For Arvind Kulkarni, the journey was more than a professional achievement. It was a testament to the power of decisive leadership, teamwork, and a commitment to excellence. Stellar Aerospace's robust CMS was not just a response to an audit finding but a foundation for sustained success in a competitive industry.

## CS 38 [AS9100D Clause 8.1.2] Configuration Status Accounting Is Not Effective

In the vibrant city of Bangalore, India, a medium-scale manufacturing unit, Stellar Aerospace Components Pvt. Ltd., was striving to solidify its presence in the aerospace sector. Known for its precision engineering, the company had set its sights on becoming a trusted supplier of aerospace parts to defense customers. However, a recertification audit revealed a critical gap that threatened their ambitions.

### The NC

During the recertification audit, the Planning Manager, Arvind Kulkarni, was again at the helm as the process owner. The auditor's meticulous review uncovered a major NC: "Configuration status accounting (CSA) is not effective when verified with customer requirements for the aircraft (standard operating procedure, SOP); deficiencies are not covered by appropriate authorization/concession from customer/regulatory authority." The auditor recorded appropriate evidence below the NC.

This revelation was a significant setback. Arvind understood the gravity of the situation. Without addressing this issue, Stellar Aerospace's path to supplying parts to defense customers would remain blocked. He resolved to transform the company's CMS to meet the highest standards.

## The Action Plan

Arvind called a meeting with his team, including representatives from marketing, engineering, production, and quality assurance. He laid out the challenge and the roadmap to overcome it.

1. Understanding the Root Cause:
   - Conduct a detailed RCA to identify why CSA was ineffective.
   - Map out gaps in the existing process, especially in linking customer requirements to the configuration of delivered parts.
2. Training and Awareness:
   - Organize workshops for all stakeholders on ISO 10007 and the principles of configuration management.
   - Collaborate with a consultant to ensure alignment with aerospace industry standards.
3. Process Overhaul:
   - Develop a CMP that includes:
     - Capturing customer requirements accurately through marketing and engineering collaboration.
     - Assigning unique configuration baselines and tracking changes throughout the product lifecycle.
     - Establishing a robust CSA system to maintain real-time traceability.
   - Introduce mandatory review and authorization of deviations by the customer or regulatory authority.
4. Implementation and Monitoring:
   - Integrate the enhanced CMP into the company's operations.
   - Conduct regular configuration audits to verify compliance with customer specifications.
   - Document and report all NCs and corrective actions in the NCR format.

### Implementation

The journey began with comprehensive training sessions that brought all teams up to speed. The marketing team adopted new templates to ensure customer requirements were captured with clarity and precision. Engineering teams implemented a system to assign unique identifiers to each configuration baseline.

The CSA system was revamped using advanced software, enabling real-time tracking of configurations. Deviations from the baseline were flagged automatically, prompting immediate action. A process was established to seek authorization or concessions from the customer or regulatory authority for any deficiencies.

Arvind also formed a CCB that met weekly to review and approve changes. Internal audits were conducted to test the new system, and feedback was incorporated to refine the processes further.

### The Follow-Up Audit

Six months later, the auditor returned to verify the corrective actions. Arvind presented comprehensive evidence:

- Training records and updated process documents.
- Reports from the CSA system show the traceability of all parts delivered.
- Documentation of customer authorizations for any deviations.
- Records of internal configuration audits and their findings (plan and execution).

The auditor meticulously reviewed the evidence and verified the system's functionality online. Satisfied with the actions taken, the auditor remarked, "The process has been significantly improved. Initial certification is recommended."

### The Road Ahead

With the recertification secured, Stellar Aerospace Components resumed its journey toward becoming a trusted supplier for defense aerospace customers. The new CMS not only ensured compliance but also became a cornerstone of the company's reputation for reliability and excellence.

A year later, during the surveillance audit, the system's effectiveness was confirmed. The auditor's report highlighted Stellar Aerospace as an exemplary case of continual improvement.

For Arvind Kulkarni, this journey was a testament to resilience, leadership, and the collective effort of a dedicated team. Stellar Aerospace's robust CMS now stands as a benchmark in the industry, paving the way for sustained success.

## CS 39 [AS9100D Clause 8.1.2] The Accidental Oversight: Experts Turning Gaps into Growth

In the serene setting of Karuizawa, a renowned customer in the aerospace industry had developed a best practice that was a source of pride and inspiration. The customer's component drawings were a marvel of precision, outlining 18 sequential process steps from raw material to finished product. These drawings not only specified the final configuration of the part but also carried the part number and drawing number with revision status, serving as a guiding beacon for their supply chain.

At the Southeast Asian (SEA) manufacturing facility, the process was seamless in many respects. The customer provided traceable raw materials, and the SEA team ensured production was well documented through a quality plan and traveler. When the parts were dispatched, they were accompanied by documentation indicating the part number, revision status, and traceability. However, during a value-added audit, a potential gap was uncovered in the delivery process.

The lead auditor, Satoshi Takahashi, was initially impressed by the meticulous adherence to the process. He praised the SEA factory for integrating the 18 process steps directly into the component drawing, which showcased their commitment to clarity and precision. However, as he examined the traveler, he noticed a critical discrepancy. The traveler recorded the completion of only 12 operations, leaving the finishing operations unmarked.

When Takahashi inquired, the production manager acknowledged the oversight. It was revealed that the part—intended to be delivered as per the drawing—had only undergone operations up to step 12. Despite this, the part was inspected, accepted, and shipped to the customer as though it were a finished product. The dispatch documentation reflected the final part number and revision status, creating a misleading impression.

"This part does not meet the complete configuration," Takahashi explained, his tone firm but constructive. "You have effectively delivered a semi-finished product while labeling it as a completed one. This is a significant NC."

The SEA team clarified that the shipment aligned with the PO and work order, which explicitly called for process steps up to step 12. However, Takahashi pointed out that the documentation—particularly the indication of the final part number—did not accurately reflect the state of the part.

A collaborative meeting followed, involving representatives from both the SEA factory and the Karuizawa headquarters. The document titled "Controlling the Interface Between KZ and SEA Factory" was reviewed, and both parties agreed that a gap in communication and documentation had led to confusion. With mutual understanding, they welcomed the NCR issued by Takahashi as an OFI.

Corrective actions were swiftly implemented. The production and dispatch documentation were revised to clearly differentiate between semi-finished and finished parts. The traveler now explicitly indicated the process steps completed at the time of dispatch, ensuring transparency and accuracy.

When the corrective actions were revisited during a follow-up audit, Takahashi found them effective. The SEA factory had not only addressed the immediate issue but had also strengthened its system to prevent similar discrepancies in the future.

"This is how an organization drives toward excellence," Takahashi concluded during the audit's closing meeting. "It is not just about addressing NCs but about using them as opportunities to refine processes and build trust."

The story of the SEA factory and Karuizawa customers became a case study of how audits can add value by uncovering hidden gaps and driving systemic improvements. It underscored the importance of clear documentation, accurate communication, and the relentless pursuit of excellence in every step of the production and delivery process.

## CS 40 [AS9100D Clause 8.1.3] Safety First, Quality Always: Stellar Aerospace's Culture of Excellence

In the vibrant industrial landscape of Bangalore, India, Stellar Aerospace Components Pvt. Ltd., a medium-scale manufacturer, was at a pivotal juncture. With aspirations to become a trusted supplier of aerospace parts to defense customers, the company faced a recertification audit critical to its ambitions. Leading the charge was Arvind Kulkarni, the diligent and forward-thinking head of planning.

### The NC: A Wake-Up Call

*During the audit, a significant NC was identified: "8.1.3: Actions to ensure product safety are not clearly defined." The auditor had recorded appropriate evidence below the NC.*

The finding cast a spotlight on a gap in the company's processes. Without clearly defined actions to ensure product safety, Stellar Aerospace risked jeopardizing its certification and, more importantly, the trust of its potential defense customers. Arvind knew that addressing this issue was not just about compliance; it was about embedding a culture where safety—both personal and product—follows where quality leads.

### The Vision: Robust Configuration Management and Safety Integration

Arvind envisioned a robust CMS that spanned the lifecycle of aero parts—from marketing and capturing customer requirements to production, configuration audits, and final delivery. This system would ensure that every part delivered matched the customer's specifications precisely and adhered to stringent safety standards.

### The Action Plan

To address the NC and elevate the company's processes, Arvind initiated a comprehensive action plan:

1. Enhancing Understanding of Product Safety:
    - Conducted training sessions for all employees on the principles of product safety in aerospace manufacturing.
    - Highlighted the best practices for ensuring safety throughout the product lifecycle.
2. Incorporating Product Safety into Configuration Management:
    - Updated the CMP to include detailed protocols for product safety at every stage of the lifecycle.
    - Established a clear process for identifying, assessing, and mitigating risks to product safety.
3. Strengthening Marketing's Role:
    - Revised marketing templates to capture detailed customer requirements, including safety specifications and standards.
    - Ensured seamless communication between marketing, engineering, and production teams.
4. Practical Actions to Ensure Product Safety in the Factory:
    - Implement rigorous inspection protocols for incoming materials to verify compliance with safety standards.
    - Conduct quality checks at each production stage to identify and resolve defects early.
    - Use traceability systems to ensure that each component's origin and processing history are documented.
    - Establish a clear process for managing and approving deviations, ensuring they meet safety standards.

- Create and enforce a final product verification checklist before delivery to customers.

5. Practical Actions to Ensure Personnel Safety in the Factory:
   - Place safety signage and instructions prominently in all work areas.
   - Conduct regular training sessions on personal protective equipment usage and emergency protocols.
   - Implement a system for identifying and mitigating workplace hazards, such as machinery maintenance checks.
   - Introduce safety barriers and clearly marked walkways to separate personnel from operational equipment.
   - Ensure proper ventilation and lighting throughout the factory.
   - Conduct periodic fire and evacuation drills to prepare employees for emergencies.
   - Monitor compliance with safety policies and take corrective actions promptly when violations occur.
6. Configuration and Product Safety Audits:
   - Developed an internal audit process to verify that parts met both configuration and safety requirements before delivery.
   - Introduced a product safety review board (PSRB) to oversee safety measures and authorize concessions or deviations.
7. Documentation and Monitoring:
   - Maintained detailed records of all safety measures, audits, and corrective actions.
   - Updated the NCR format to include product safety-specific metrics.

## Implementation and Success

The cross-functional team executed the plan with precision. Employees were empowered through training, and the factory floor became a model of safety and efficiency. Marketing teams captured customer requirements more effectively, and engineering teams ensured these were integrated into designs. Configuration audits included rigorous safety checks, guaranteeing that the delivered parts met all specifications and safety standards.

## The Follow-Up Audit: A Turning Point

Six months later, the auditor returned to assess the corrective actions. Arvind presented extensive evidence, including:

- Training records and updated CMP documentation.
- Reports from safety and configuration audits.
- Minutes from PSRB meetings detailing safety decisions.
- Photographs and videos of the revamped factory setup, showcasing safety measures.

After reviewing the evidence and conducting online verifications, the auditor remarked, "Stellar Aerospace has transformed its approach to product safety. Initial certification is recommended."

## The Road Ahead

With certification secured, Stellar Aerospace was ready to take on the defense aerospace market. The company's robust CMS, underpinned by a strong safety culture, became its competitive edge. During the next surveillance audit a year later, the auditor confirmed the effectiveness of the implemented actions, cementing Stellar Aerospace's reputation as a benchmark for quality and safety.

For Arvind Kulkarni, this achievement was a milestone in his career and a testament to the belief that "safety follows where quality leads." Stellar Aerospace's systems now stand as a beacon of excellence in the aerospace industry, inspiring other manufacturers to aim higher.

## CS 41 [AS9100D Clause 8.1.4] Planning Head's Resolve: Guarding Against Counterfeits

In the bustling industrial unit of an aerospace manufacturing company in Goa, India, an initial certification audit was underway. The process was grueling; every corner of the organization was meticulously examined. The goal was not just certification but setting the stage for supplying aerospace parts to defense customers—a critical milestone for the company.

The planning department, under the leadership of the Head of Planning, Arjun Menon, found itself under the spotlight. Known for his strategic foresight and uncompromising stance on quality, Arjun was prepared for the challenges ahead.

*However, the audit brought forth a significant NC: "Actions to prevent the presence of counterfeit parts in the final product were not clearly defined or verified." Clause 8.1.4. The auditor had recorded appropriate evidence below the NC.*

This finding hit at the heart of aerospace manufacturing—where quality and safety are paramount. Counterfeit or unauthorized parts could jeopardize not only the company's reputation but also the lives of countless individuals. Arjun recognized that the issue demanded immediate, decisive action.

### RCA

The audit revealed that while the company had systems in place for part approval and traceability, the measures to prevent counterfeit parts from entering the supply chain were fragmented. Suppliers occasionally bypassed stringent checks, and the process to isolate and control suspect parts lacked robustness.

Arjun convened a meeting with his core team, including quality engineers, procurement officers, and legal advisors. "This is not just about passing an audit," he emphasized. "This is about safeguarding the integrity of our products and ensuring we become a trusted partner for defense clients."

Arjun brought out the following information for further study and actions:

1. Counterfeit Parts Prevention Strategies Guide prepared by David C. Meshel, Joint Space Protection GEOINT Futures Office (AEROSPACE REPORT NO. TOR-2014-02200) Prepared for: National Reconnaissance Office 14675 Lee Road Chantilly, VA 20151-1715 Authorized by: National Systems Group (approved for public release; distribution unlimited) is a good reference document in this regard.
2. A draft work instruction is also available in Para 8.1.4 Prevention of Counterfeit Parts, in the book "Implement AS 9100 D/EN 9100:2018 for Business Excellence - Quality Management Systems Requirements for Aviation, Space and Defence Organisations (Same as EN 9100:2016; Includes ISO 9001:2015).

### Building the Counterfeit Prevention Framework

With the root causes identified, Arjun spearheaded the development of a **counterfeit prevention framework**. This comprehensive system encompassed every stage of the production lifecycle, from marketing to delivery:

- **Marketing and Configuration Capture**: The marketing team was trained to collaborate closely with customers to capture precise configurations. Detailed drawings and specifications were validated and stored in a secure, centralized system accessible only to authorized personnel.

- **Supplier Onboarding, Approvals, and Regular Performance Monitoring**: Arjun insisted on stricter supplier qualification processes. Every supplier had to provide verifiable certifications, including traceability documentation for all parts supplied. Noncompliance with contract requirements with respect to supplying counterfeit parts resulted in immediate disqualification and, if necessary, legal action.
- **Incoming Inspection and Control**: Advanced inspection protocols were introduced. Each part was checked against approved configurations using state-of-the-art testing equipment. Suspect parts were quarantined and marked to prevent re-entry into the supply chain. Information was sent to the legal authority for taking necessary actions.
- **Production Line Safeguards**: An automated tracking system was integrated into the assembly line. This system flagged any part that deviated from the approved configuration, ensuring only authorized and approved parts were used.
- **Training and Awareness**: Regular training sessions were conducted for employees at all levels. The focus was on recognizing counterfeit parts, understanding the importance of traceability, and adhering to stringent quality standards to ensure prevention, detection, and withdrawal of counterfeit or suspected counterfeit parts found in the production line.
- **Regular Audits and Reviews**: A schedule for internal audits was established, ensuring continual monitoring of the system's effectiveness. Arjun personally reviewed audit findings to maintain accountability.

## Closing the NC

The corrective actions were documented in the NCR and presented to the auditor. Evidence included revised procedures, supplier certification records, training logs, and test reports from the upgraded inspection processes.

The auditor, impressed by the thoroughness and speed of implementation, accepted the corrective actions. "This framework is not just a response to the NC but a benchmark for the industry," the auditor remarked.

## The Road Ahead

The company received its initial certification, a testament to the dedication and leadership of Arjun and his team. The auditor planned to verify the effectiveness of the corrective actions in the next audit, scheduled for the following year.

Meanwhile, Arjun did not rest on this success. He continued refining the counterfeit prevention system, collaborating with industry experts and defense agencies to ensure the company met and exceeded global standards.

Under his stewardship, the industrial unit was no longer just a manufacturing hub. It became a symbol of excellence, ready to deliver world-class aerospace parts to defense customers, where every bolt, rivet, and component guaranteed quality and authenticity.

## CS 42 [AS9100D Clause 8.7.1] When the Marketing Chief Handled Customer Complaints

The auditor was reporting the serious NC to the MD: "The organization was not taking actions necessary to contain the effect of the nonconforming product identified by a customer."

You may be wondering how audits sometimes become a company's great savior. Let me tell you a story about how a major audit finding the auditor reported to the MD saved an organization from disaster!

### The Turning Point: How an Audit Saved the Company

Sunbeam Aerospace Solutions, a South India-based organization specializing in precision manufacturing for the aviation sector, prided itself on its robust operations and impressive annual business volume of ₹250 crores. However, lurking beneath the surface of this success was a systemic issue—customer complaints and product rejections, totaling 5%, were being ignored.

The company's "marketing process" chief, a seasoned professional with years of experience, was confident in his department's approach—or lack thereof. "We can afford that rejection rate," he told the auditor during an AS9100D audit, adding, "We still earn a good profit. We do not do anything in particular to address complaints or reduce rejections."

The auditor, Srinivasan, listened intently, probing deeper.

### The Audit Findings

**Auditor:** *"What is your annual business volume?"*

**Marketing Chief:** *"₹250 crores."*

**Auditor:** *"At a 5% rejection rate, the cost of poor quality is more than ₹12.5 crores annually. Imagine what could be achieved if rejections were reduced to just 1%—a savings of ₹12.5 crores annually."*

The room fell silent. The implications were clear, but the marketing chief brushed it off with a smile.

When questioned about corrective actions for customer complaints, the chief admitted:

- Human errors were noted as reasons for rejection, but no corrective or preventive actions were taken.
- No record of steps taken to address complaints existed.

Management Representative Srinivasan summarized the issue in his report to senior management:

1. The organization lacked a structured approach to handling customer complaints and rejections.
2. The cost of poor quality was being ignored under the guise of profitability.
3. No system was in place to address root causes or ensure continual improvement.

### The Auditor's Challenge

During a working lunch with the leadership team, Srinivasan posed a challenge:

*"I will give an expert ₹1 crore, one year, and the authority to implement a corrective system. Can he reduce the rejections from 5% to 1%?"*

*The auditor added: "This will not only recover ₹11.5 crores in annual losses but also create a sustainable improvement framework for years to come."*

The idea left the room stunned. The Chairman, Raghavan, realized the magnitude of what was at stake—not just the financial losses, but also the company's reputation in the highly competitive aerospace sector.

## The Aftermath

The audit findings were a wake-up call. The board of directors convened an emergency meeting, and sweeping changes followed:

1. Restructuring Leadership:
    - The Marketing Chief, along with 50 members of his team, was relieved of their duties.
    - This decision was based on their refusal to address customer complaints and their complacency in maintaining the status quo.
2. Transparency from Management:
    - The Chairman acknowledged that the rejection rate was "affordable" because it was factored into the pricing. However, this approach masked inefficiencies and created a false sense of profitability.
    - Management admitted that the "quality system was effective on paper" but failed in execution.
3. Appointing a Quality Expert:
    - A seasoned quality improvement expert was hired with the promised ₹1 crore budget and authority to revamp the system.
    - The expert implemented a robust corrective action system, focusing on:
        - RCA for each rejection and complaint.
        - Revising processes to eliminate human errors.
        - Establishing a customer feedback loop for continual improvement.
4. Results Within a Year:
    - The rejection rate dropped from 5 to 1%, saving ₹11.5 crores annually.
    - Customer satisfaction improved, leading to an influx of new business.

## A Lesson for the Board

At the annual review meeting, the Chairman addressed the board:

***Chairman:***

*"The audit was a turning point. It revealed not just flaws in our processes but also complacency in our leadership. By ignoring customer complaints and tolerating rejections, we risked our reputation, profitability, and future growth. Thanks to the audit, we've turned a blind spot into an opportunity for transformation. Today, Sunbeam Aerospace is stronger than ever, with a commitment to quality that matches the standards of the industry we serve."*

The board applauded the changes, recognizing the vital role audits play—not just as a compliance exercise, but as a tool for introspection and growth.

## Conclusion

This story highlights how a well-conducted audit can uncover systemic issues and drive transformative change. For Sunbeam Aerospace, the audit didn't just save ₹11.5 crores annually—it saved the company itself.

## CS 43 [AS9100D Clause 8.2.2] Mind the Gap: Fixing Customer Requirement Oversights

SkyTech Systems, a growing engineering company, was on the verge of entering the ASD market. To achieve this, the organization sought AS9100D certification to demonstrate its capability to meet the exacting demands of this industry. Among the various audits being conducted, the marketing and business development processes were under scrutiny.

James Carter, the experienced Business Development Manager, was confident as he sat across the table from the auditor, Amanda Blake. He was prepared to showcase SkyTech's customer acquisition strategy and order management processes. However, as the discussion unfolded, James found himself facing an unexpected challenge.

### The Audit Discussion

Amanda began by reviewing how SkyTech acquired customer requirements for their products and services. She asked, "James, during the order intake process, how does your team determine the requirements for the products and services to be offered? Specifically, do you ensure the following:

1. The **special requirements** of the product are identified.
2. The **operational risks**—like using new technology, capacity to deliver, or meeting tight deadlines—are considered."

James hesitated. "Well, we discuss the general requirements with the customer and pass the details to the design team for further review."

Amanda raised her eyebrows. "That's a start, but how do you ensure these specific considerations are addressed during the initial stages of customer engagement?"

James, caught off guard, admitted, "I'm not sure. Could you explain how we could determine those aspects?"

Amanda shook her head gently. "I'd love to, but my role as an auditor is to assess your current process, not guide you through the implementation. My recommendation is to study the AS9100D standard, especially clause 8.2.1, along with related materials like AS9103. Additionally, collaborate with your design and engineering teams—they can provide insight into identifying special requirements and operational risks."

### The NC

Amanda issued the NC: **clause 8.2.2—when determining customer requirements, the organization has not ensured that:**

- **Special requirements** of products and services are determined. It included key characteristics and critical items.
- **Operational risks** (e.g., new technology, ability and capacity to provide, short delivery timeframe) are identified.

Amanda concluded, "Take corrective actions to address this gap. I'll return to verify your solutions. Once resolved, I'll be happy to recommend your certification."

### The Response and Follow-Up Actions

James, eager to close the gap, convened a meeting with the marketing team and collaborated with the design and engineering departments. Together, they devised a structured plan:

1. Containment Action:
    - Immediately implement a checklist for customer interactions to capture special requirements and potential operational risks during order discussions.
    - Review all ongoing projects to identify any overlooked requirements or risks.
2. RCA:
    - Investigate why special requirements and operational risks were not being addressed.
    - James identified a lack of awareness about the AS9100D requirements and limited cross-functional collaboration during order intake.
3. Corrective Actions:
    - **Develop a Customer Requirements Checklist:** Include fields to identify special requirements (e.g., performance tolerances, regulatory needs) and potential operational risks (e.g., production capacity, new technologies).
    - **Training Sessions:** Educate the business development and marketing teams on AS9100D clause 8.2.2 and related standards, including real-world examples from the design team.
    - **Cross-Functional Review Process:** Establish a protocol where every customer requirement is reviewed by design, quality, and operations teams to identify risks and validate special requirements.
    - **Integration into CRM:** Modify the customer relationship management (CRM) system to prompt the business development team to capture these elements during order entry.
4. Verification:
    - Conduct mock customer order reviews to validate the effectiveness of the new process.
    - Create a log of identified special requirements and risks, and document how these were managed.

### The Outcome

Two months later, Amanda returned for a follow-up audit. James proudly presented the newly implemented process, including the updated CRM system, training logs, and sample customer requirement reviews.

Amanda was impressed. "This is a significant improvement. You've not only addressed the NC but also enhanced your approach to capturing customer needs. It is clear that you're ready for the next steps in AS9100D certification."

For James, the experience was transformative. He reflected, "This was a learning curve for me and my team. Understanding customer requirements is not just about getting the job done—it is about ensuring we deliver precisely what's needed while minimizing risks. This has prepared us for the rigor of the ASD market."

With the NC resolved and its processes stronger than ever, SkyTech Systems was one step closer to achieving its certification and entering a new era of growth.

## CS 44 [AS9100D Clause 8.2.2] Stamped But Not Sealed: The Journey to Customer Alignment

SkyTech Systems had faced its share of challenges on its journey toward AS9100D certification. A year earlier, during an internal audit, a critical gap had been identified in the marketing and order management processes. This NC highlighted a fundamental weakness in how customer and product requirements were reviewed, both before quotations and after order acceptance.

### The Original NC

The auditor, Sarah, had raised a red flag during her audit. While reviewing the marketing process, she noticed a troubling pattern. Customer orders were being processed with a simple "order acceptance" stamp, and there was no evidence of a thorough review to consolidate or validate the requirements. Sarah asked the marketing manager, "Can you show me how SkyTech determines and reviews customer and product requirements before quoting and after order receipt?"

The marketing manager, unsure, replied, "We rely on the sales team's discussions with the customer. Once we get the order, we stamp it for acceptance and move it to production."

Sarah frowned. "But how do you ensure that the customer's requirements, including technical and operational aspects, are fully understood and feasible? Is there a cross-functional team review to validate requirements at both stages?"

The answer was a silent shrug.

Sarah issued the NC:

*Clause 8.2.2—determination of customer/product requirements and their review are not consolidated/established. No cross-functional team review evidenced. Reviews were not done before the quotation and again after receipt of the order.*

She explained, "Without a structured review process, you risk accepting orders that are misaligned with your capabilities or fail to meet customer expectations. This gap must be addressed immediately to ensure compliance and operational effectiveness."

### Immediate Actions Taken

SkyTech responded swiftly. Under the leadership of the business development and marketing teams, the organization undertook corrective actions to close the gap:

1. Containment Action:
   - Immediately halted the issuance of quotations and acceptance of new orders without a documented review process.
   - Created a temporary review checklist to guide the sales team during customer interactions.
2. RCA:
   - Identified that the reliance on informal communication and lack of a cross-functional review stemmed from a siloed approach to marketing, sales, and operations.
   - There was no awareness of the AS9100D requirement for consolidated reviews at critical stages.

3. Corrective Actions:
   - **Establish a Review Process:** Developed a two-step review system:
     - **Pre-Quotation Review:** Conducted by a cross-functional team, including representatives from sales, engineering, quality, and production, to validate customer requirements, special considerations, and operational feasibility.
     - **Post-Order Review:** A second review after order receipt to ensure all requirements are documented, clarified, and feasible for production.
   - **Cross-Functional collaboration:** Instituted regular meetings where teams discussed requirements and shared feedback before proceeding.
   - **Standardized Documentation:** Introduced a customer requirements review form to capture all critical information, including product specifications, special requirements, and risks, for both stages.
   - **Training Programs:** Conducted workshops to educate the sales and marketing teams on AS9100D requirements and the importance of thorough reviews.
4. Monitoring and Continual Improvement:
   - Set up periodic audits to ensure compliance with the new review process.
   - Incorporated the review process as a key metric in management review meetings.

### A Year Later

One year after the NC was issued, another audit was conducted to verify compliance. Sarah returned to SkyTech to assess the progress.

As she reviewed the updated processes, Sarah observed a significant transformation. Pre-quotation and post-order reviews were now embedded into the workflow. The cross-functional team review meetings were well documented, and the new customer requirements review form provided clear evidence of validation at each stage.

The marketing manager proudly explained, "We've come a long way. The pre-quotation review helps us avoid overpromising, and the post-order review ensures we deliver exactly what the customer needs. These processes have improved our customer satisfaction and internal efficiency."

### The Outcome

Sarah was impressed. "You've addressed the NC effectively. Your processes now align with AS9100D requirements, and you've built a system that will support your growth in the aerospace and defense sectors."

For SkyTech, the experience was a valuable lesson in the importance of robust processes and cross-functional collaboration. The NC had not only driven compliance but had also laid the foundation for greater organizational efficiency and customer trust.

## CS 45 [AS9100D Clause 8.2.3.2] When the Records Vanished: PrecisionTech's Wake-Up Call

In the bustling industrial zone of Singapore, PrecisionTech Engineering, a built-to-print organization, was undergoing its routine audit for AS9100 compliance. The company prided itself on delivering high-quality products, built exactly to customer specifications. However, as the external auditor Khadgi Singh began her meticulous review, a concerning gap emerged.

During the documentation audit, Khadgi asked to see the records of customer contract reviews for several recent jobs given by the customers. James, the company's Management Representative, hesitated.

"We do not maintain formal review records," he admitted. "Our team discusses the designs during meetings, but we've never documented those reviews."

Khadgi frowned but maintained his professional demeanor. "James, documenting and retaining records of contract reviews is a critical requirement. It ensures traceability, accountability, and evidence that due diligence has been exercised in identifying and addressing customer requirements and product requirements and associated risks."

She issued a NC against the missing review records and explained the implications. "Without documented evidence, it is impossible to verify whether the reviews were conducted effectively. This could lead to potential quality issues or customer dissatisfaction down the line."

Realizing the seriousness of the issue, James gathered his team to address the gap. Over the next week, they:

1. Implemented a structured contract review process.
2. Created a standard template to document the date, participants, key discussions, and decisions of each review.
3. Reviewed past projects to retrospectively create records for critical contracts, based on available meeting notes and recollections.
4. Make a consolidated checklist for verification before sending quotations and after receiving customer orders.

Once the corrective actions were in place, PrecisionTech invited Khadgi back for a follow-up audit. She carefully examined the newly implemented process and reviewed the documented records of recent contract reviews.

"This is a significant improvement," Khadgi said with a nod. "You've established a robust system to ensure contract reviews are documented and retained. Going forward, this will enhance both compliance and quality assurance."

James felt a wave of relief and pride as Khadgi closed the NC. The experience was a wake-up call for the organization, reinforcing the importance of not only adhering to standards but also fostering a culture of accountability and continual improvement.

From that point on, PrecisionTech's contract review records became an integral part of their quality system—a testament to their commitment to excellence.

## CS 46 [AS9100D Clause 8.2.3.2] Missing Logistics: AeroPrecision's Path to Accuracy

In Singapore's advanced aerospace manufacturing district, AeroPrecision Systems Pvt. Ltd., a company specializing in built-to-print aerospace components, was undergoing an AS9100D audit. Known for its high-quality work and timely deliveries, AeroPrecision had been eager to achieve certification to enhance its reputation in the competitive aerospace industry.

The lead auditor, Amira Tan, began her assessment with a focus on contract reviews, particularly how the company documented customer requirements, including logistics considerations.

"Can you show me the records for the contract reviews of your last few projects?" Amira asked as she sat across from Benjamin Lim, the Management Representative.

Benjamin retrieved a stack of documents but hesitated as Amira flipped through them. Her brows furrowed.

"Benjamin," Amira said, her tone professional but firm, "I see no documented information on how logistics requirements were reviewed or agreed upon. AS9100D clause 8.2.3.2 explicitly requires retaining records of contract reviews, including any new product or service requirements, and that includes logistics like delivery timelines, packaging needs, and shipping methods."

Benjamin nodded, realizing the gap. "We do discuss logistics during meetings and emails with clients, but we haven't been documenting these reviews formally. Our focus has been on technical specifications rather than the full scope of requirements."

*Amira issued an* ***NC****, explaining its significance. "In aerospace, logistics is as critical as the product itself. Delayed or improperly packaged components can lead to supply chain disruptions. Without documented reviews, there is no assurance that logistics requirements have been fully understood or met."*

### The Corrective Actions

Determined to close the gap, Benjamin led the team in addressing the issue:

1. **Formalizing the Review Process:** A comprehensive review checklist was developed to ensure that logistics requirements, including timelines, packaging standards, and shipping methods, were thoroughly reviewed during contract discussions.
2. **Documentation Templates:** A standard template for contract review records was introduced, ensuring all details—technical specifications, product requirements, and logistics—were documented and approved by both parties, including any changed requirements from the customers.

3. **ERP Integration:** The new review process was integrated into the company's ERP system to centralize records and streamline updates.
4. **Training and Awareness:** The sales, logistics, and project teams were trained to implement the updated process effectively, ensuring consistent documentation.

### The Follow-Up Audit

Two weeks later, Amira returned for the follow-up audit. Benjamin confidently presented the new process and records for several recent projects, including detailed documentation of logistics requirements.

Amira reviewed the documents carefully. "This is excellent work, Benjamin," she said. "You've addressed the NC comprehensively. Your new system not only meets AS9100D requirements but also enhances traceability and accountability."

With the NC closed, AeroPrecision passed the audit phase successfully. Benjamin felt a sense of pride, knowing his team had turned a compliance gap into an OFI.

From that day forward, AeroPrecision's meticulous approach to documenting contract reviews became a cornerstone of their operations, ensuring they met both customer expectations and the high standards of the aerospace industry.

## CS 47 [AS9100D Clause 8.2.4] The Cost of Miscommunication: Managing Changing Customer Requirements

In the highly competitive aerospace manufacturing sector, the Singapore-based firm prided itself on its customer-first approach. This ethos had secured them high-profile clients, but during an annual **AS9100D surveillance audit**, their customer-centric mindset was revealed as both a strength and a vulnerability.

### The NC

*The audit highlighted a critical NC under **clause 8.2.4**: "The organization shall ensure that relevant documented information is amended and that relevant persons are made aware of the changed requirements when the requirements for products and services are changed. This has not happened when the customer changed their requirements after the placing of an order." Appropriate evidence was recorded by the auditor below the NC.*

This oversight had far-reaching consequences. The absence of a systematic process to handle changing customer requirements led to inefficiencies, misaligned production efforts, and significant financial losses.

### The Managerial Misstep

The management team, including Ravi Kumar, the Operations Manager, believed they were doing the right thing by prioritizing customer satisfaction. When a key client requested design changes after placing an order, the team immediately halted production and made adjustments without fully evaluating the implications.

The result was a series of cascading problems:

1. **Production Delays:** Restarting production disrupted the workflow, affecting other orders in the pipeline.
2. **Increased Costs:** Scrap material, overtime wages, and expedited logistics to meet deadlines eroded profitability.
3. **Employee Frustration:** The lack of clear communication led to confusion and inefficiencies on the shop floor.

"We thought we were pleasing the customer," Kumar admitted. "But we ended up hurting the company."

## The Process Owner Steps in

The NC caught the attention of the process owner, Noor Amina, a meticulous and pragmatic professional from Malaysia. Known for her ability to balance customer needs with operational efficiency, Noor recognized that this issue was more than a compliance failure—it was a threat to the company's sustainability.

"This is not just about satisfying the auditor," she told her team. "It is about ensuring we can deliver quality products to our customers without jeopardizing our financial health or our people's morale."

## Containment Actions

To address the immediate risks:

1. Noor instituted a **temporary freeze** on accepting any mid-order changes without managerial approval.
2. A task force was formed to identify all orders affected by recent customer changes and ensure proper communication with the production team.
3. Customers were notified about potential delays caused by the adjustments, and new delivery schedules were negotiated.

## RCA

Noor facilitated an RCA with input from cross-functional teams. The findings were:

1. **Lack of Documentation Updates:** Changes to customer requirements were not systematically documented or communicated to the relevant team.
2. **Absence of Change Management Protocols:** There was no formal process to evaluate the impact of changes before implementing them.
3. **Customer-Driven Decisions:** A reactive approach to pleasing customers often leads to hasty, costly decisions.

## Corrective Actions

Noor developed a robust corrective action plan to address these issues:

1. Formalized Change Management Process:
    - A structured process was introduced to handle customer-requested changes, including:
        - A mandatory impact assessment before accepting changes.
        - Documentation updates through a central system accessible to all relevant personnel.
    - Changes were categorized based on their complexity and urgency, ensuring a proportionate response.

2. Training and Awareness Programs:
   - Employees were trained on the new change management process and the importance of adhering to AS9100D standards.
   - Customer service teams were educated on how to communicate realistic expectations to clients.
3. Cost–Benefit Analysis Requirement:
   - A cost–benefit analysis was made mandatory for every change request, enabling the management to make informed decisions about whether to accept, negotiate, or decline changes.
4. Enhanced Customer Engagement:
   - Customers were encouraged to finalize requirements during the initial stages of order placement.
   - Regular checkpoints were introduced to align expectations and mitigate late-stage changes.

### Reporting and Auditor Feedback

The corrective actions, supported by evidence such as updated procedures, training records, and examples of implemented changes, were reported to **OASIS**. The auditor reviewed and accepted the corrective action plan, commending the firm for its comprehensive approach.

However, the auditor emphasized that the effectiveness of these measures would be scrutinized during the next annual audit.

### Turning a Challenge into an Opportunity

The changes implemented by the firm not only resolved the NC but also improved operational efficiency and customer satisfaction. One notable client remarked, "Your new process gives us greater confidence. Knowing that every change is evaluated thoroughly reassures us about your commitment to quality and delivery timelines."

The cost–benefit analysis became a powerful tool for convincing customers to limit unnecessary changes, thereby protecting both parties from potential delays and cost overruns.

### A Year Later

When the auditor returned for the follow-up audit, they found a transformed organization. The change management process was seamlessly integrated into operations, and the firm avoided any significant losses due to mid-order changes over the past year.

"Your ability to learn from setbacks is remarkable," the auditor noted. The firm not only secured its **AS9100D certification** but also strengthened its reputation as a reliable and customer-focused aerospace manufacturer.

For the company, the experience underscored a vital lesson: aligning customer satisfaction with operational excellence is the key to long-term success.

## CS 48 [AS9100D Clause 8.2.4] The Costly Oversight

In Bengaluru, India, TechNova Solutions, a renowned giant in software development, was in the news for an unfortunate reason. Its chairman, Arjun Desai, recently made headlines after revealing that the company suffered significant financial losses due to repeated changes in requirements from a government department client. Despite having completed half the work, the team had to redo everything multiple times, leading to delays and skyrocketing costs.

As fate would have it, the company's AS9100D surveillance audit was scheduled shortly after this controversy. The lead auditor, Devendran Murthy, entered the company premises fully aware of the public statement but approached the audit with an unbiased perspective.

### Discovering the NC

While auditing one of the government projects, Devendran asked to review the contract, project scope, and change management records. He spoke with the project manager, Nidhi Verma, to understand how the company managed customer-requested changes.

"Well, the government client kept introducing new requirements after the initial agreement," Nidhi explained. "We had already completed significant portions of the work, but we had to redo it multiple times to accommodate their changes."

"Were these changes formally reviewed, documented, and evaluated for cost and time impact?" Devendran asked.

Nidhi hesitated. "No, not really. Since it was a government project, we didn't want to push back or negotiate too hard. We just made the changes."

Devendran reviewed the contract review process and change management system and found glaring gaps. There were no records of customer-requested changes, no evaluations of their cost or time implications, and no records of rework or associated expenses.

"This is a major NC," Devendran stated. "Clause 8.2.4 of AS9100D clearly requires changes to customer or product requirements to be reviewed, evaluated, and documented. Without these records, you can't track the impact on your resources or negotiate with clients for additional costs."

### Issuing the NC and Recommendations

Devendran issued an NCR with the following observations:

- **NC:** Changes to customer/product requirements were not reviewed, nor were any records retained.
- **Impact:** The lack of documented reviews and cost assessments led to uncontrolled rework and financial losses.

He strongly advised the company to:

1. Finalize Contracts Before Starting Work:
   - Avoid redoing projects until all requirements are clearly defined and agreed upon in writing.
2. Implement a Change Management Process:
   - Develop a formal process to review, document, and evaluate the impact of changes in requirements.
   - Include cost and time evaluations for each change.
3. Negotiate Change Costs:
   - Establish a clause in contracts requiring clients to bear the cost of significant changes.
   - Engage legal and financial teams in negotiations to protect the company's interests.

4. Train Employees:
    - Conduct training for project managers and sales teams on the importance of documenting changes and negotiating costs.

### Corrective Actions and Follow-Up

The leadership team at TechNova Solutions took the feedback seriously. Over the next two months, the company implemented the following corrective actions:

- Created a robust change management framework integrated into their project management software.
- Trained employees on contract review and change documentation processes.
- Amended all future contracts to include clauses addressing change requests and associated costs.
- Retrospectively documented major changes in ongoing projects and began negotiating with the government department for partial reimbursement of the rework costs.

During the follow-up audit, Devendran reviewed the updated processes and records for a new government project. This time, all changes were thoroughly documented, their impacts evaluated, and additional costs negotiated with the client.

"This is a significant improvement," Devendran said. "You've closed the gap effectively. With these controls in place, you'll prevent further financial losses and enhance your ability to manage projects successfully."

The NC was officially closed, and TechNova Solutions turned a painful lesson into an opportunity to strengthen its operations, ensuring such costly oversights would not recur.

## CS 49 [AS9100D Clause 8.3.1] Plan It Right, So It Takes Flight: How AeroSoft Nailed AS9100D

In the bustling tech hub of Bengaluru, India, AeroSoft Designs Pvt. Ltd. has established itself as a promising player in software design verification for the aerospace industry. The company was on the verge of securing its AS9100D certification, a critical milestone that would unlock new opportunities with global aerospace clients. However, during the initial certification audit, the firm encountered a significant setback under clause 8.3.1.

### The NC

*The audit uncovered a glaring gap: **"No evidence to show that systematic planning for design and development activity has been carried out."***

*The auditor had indicated appropriate evidence with project names/numbers.*

The auditor highlighted the absence of structured **project initiation files** (PIFs) or equivalent documentation that should outline:

- Project scope.
- Customer requirements.
- Resource allocation.
- Initial contract review including special requirements and key characteristics, regulatory requirements, and contract requirements.
- Milestones and deliverables.
- Verification and validation activities in coordination with customers.
- Risk assessment and mitigation plans.

This shortfall was particularly concerning, given the company's ongoing project for a prestigious US-based aerospace customer, a leader in avionics systems.

### The Challenge

The lack of systematic planning had caused several issues during the design verification project:

1. **Ambiguity in Requirements:** Without a consolidated plan, teams often misunderstand or overlook critical customer requirements.
2. **Resource Overruns:** The absence of clear task allocation led to inefficiencies and delays.
3. **Inconsistent Quality Checks:** Design verification activities lacked a structured sequence, increasing the risk of errors.

The management team, led by Project Lead **Meera Nair**, realized the gravity of the situation. "Without a strong foundation, even the best talent cannot deliver consistently," she admitted.

### Containment Actions

To address immediate concerns:

1. Meera directed her team to compile all scattered documentation related to ongoing projects into centralized PIFs.
2. A temporary task force was established to ensure that key customer requirements were clarified and communicated to all stakeholders.
3. The customer was updated on the steps being taken to strengthen the planning process, assuring them of uninterrupted project progress.

### RCA

Ms. Nair conducted a thorough RCA with her team:

1. **Lack of a Defined Planning Framework:** There was no formal process for initiating projects, leaving teams to rely on ad-hoc methods.
2. **Over-Reliance on Expertise:** The organization assumed that experienced team members could manage without documented planning.
3. **Inadequate Training:** Employees were unaware of the importance of structured planning in meeting AS9100D requirements.

### Corrective Actions

With the auditor's feedback in mind, Meera spearheaded a systematic overhaul of the company's D&D planning process:

1. **Introduction of PIFs:** A standardized PIF template was created to document:
   - Project scope and objectives.
   - Customer requirements and deliverables.
   - Resource allocation, timelines, and milestones.
   - Risk identification and mitigation strategies.
2. **D&D Planning Checklist:** A checklist was developed to ensure every phase of the project, from initiation to delivery, was systematically planned and documented.
3. **Training Programs:** Employees were trained on the new planning framework, emphasizing its role in meeting AS9100D standards and ensuring project success.
4. **Customer Engagement:** Regular review meetings were scheduled with customers to validate requirements and align expectations.
5. **Monitoring and Review Mechanism:** A process was established to review the effectiveness of planning activities at regular intervals.

### Reporting and Auditor Feedback

The corrective actions and supporting evidence, including completed PIFs, training records, and project review reports, were uploaded to **OASIS**. The auditor acknowledged the improvements and accepted the corrective action plan.

The auditor remarked, "This level of planning not only addresses the NC but also positions your company to deliver superior results for your clients."

### Benefits Realized

The implementation of systematic planning brought tangible benefits to AeroSoft Designs Pvt. Ltd.:

- **Enhanced Efficiency:** Teams now had a clear roadmap, reducing time wasted on resolving ambiguities.
- **Improved Quality:** A structured approach ensured consistent adherence to customer requirements and project timelines.
- **Stronger Customer Relationships:** The US client praised the company for its proactive measures, leading to discussions about future collaborations.
- **Informed Decision-Making:** The risk assessment framework enabled the team to anticipate and address potential challenges early.

### A Year Later

When the auditor returned for the follow-up audit, they found a transformed organization. The design verification function was now a model of systematic planning and execution. The auditor was particularly impressed with how the changes had been seamlessly integrated into the company's culture.

"Achieving compliance is one thing; sustaining it through continual improvement is another," the auditor noted during the closing meeting.

For AeroSoft Designs Pvt. Ltd., the experience was more than just a step toward certification—it was a catalyst for operational excellence. The firm had not only secured its AS9100D certification but had also positioned itself as a trusted partner for global aerospace clients.

## CS 50 [AS9100D Clause 8.3.2] Building a Foundation: Enhancing Input Verification for Aircraft Design

In the fast-growing aerospace software sector, SkyTech Solutions Pvt. Ltd., headquartered in Hyderabad, India, was emerging as a promising name. Known for its focus on design verification, the company had been entrusted with a critical service delivery project for a prestigious US-based aerospace customer, a leader in flight control systems.

As part of their journey toward **AS9100D certification**, the company underwent its initial certification audit. However, during the audit, a significant gap was identified under **clause 8.3.2**:

> *"The adequacy of inputs was not verified and confirmed before proceeding with the service delivery." The auditor substantiated his statement with appropriate evidence (projects in hand).*

This discovery became a turning point for SkyTech Solutions, compelling the team to rethink their approach to managing project inputs and laying the groundwork for long-term improvement.

### The NC

The auditor's findings revealed a systemic issue:

- **Input Ambiguity:** Customer-provided inputs, such as system requirements, interface specifications, and performance criteria, were often incomplete or unclear.

- **Lack of Verification:** The design team assumed the accuracy of these inputs without formal verification, leading to rework and inefficiencies.
- **Impact on Delivery:** Delays and inconsistencies in meeting project milestones resulted in jeopardizing both timelines and customer satisfaction.

The process owner, Project Lead Aditya Sharma, acknowledged the oversight. "We were so eager to get started that we didn't fully validate what we were working with," he admitted during the audit's closing meeting.

### Immediate Containment

Aditya took swift action to address the issue:

1. **Freeze on New Tasks:** No further design activities were undertaken until all inputs for the ongoing project were verified.
2. **Cross-Functional Input Review:** A temporary task force, including representatives from the design, quality, and customer support teams, reviewed and validated all existing inputs.
3. **Customer Collaboration:** The client was engaged to clarify and confirm ambiguous or missing requirements.

These containment actions prevented further missteps and reassured the customer of SkyTech's commitment to quality.

### RCA

Aditya led a detailed RCA, which uncovered the following issues:

1. **Absence of Input Verification Processes:** There was no formal mechanism to validate customer-provided inputs before starting a project.
2. **Over-Reliance on Customer Documentation:** Teams assumed that customer-provided inputs were complete and accurate without verification.
3. **Lack of Communication Channels:** Inadequate communication with customers led to delays in resolving input-related queries.

### Corrective Actions

With insights from the RCA, Aditya and his team developed a comprehensive corrective action plan:

1. Implementation of Input Verification Procedures:
    - A structured process was introduced to verify and confirm the adequacy of all customer-provided inputs.
    - Key steps included:
        - Cross-functional review of inputs against project requirements.
        - Documentation of discrepancies and resolutions.
        - Formal customer approval of clarified inputs.
2. Input Validation Checklist:
    - A checklist was created to ensure that inputs met predefined criteria, such as completeness, clarity, and alignment with customer expectations.
3. Enhanced Communication Protocols:
    - A dedicated liaison team was formed to facilitate direct communication with customers for the timely resolution of input-related issues.

- Regular input review meetings with customers were instituted to ensure alignment throughout the project lifecycle.

4. Training Programs:
    - Teams were trained on the importance of input verification and new procedures, emphasizing how this aligned with AS9100D standards.

## Reporting and Auditor Feedback

The corrective actions, along with supporting evidence such as completed input verification checklists, training records, and customer approval documentation, were submitted to **OASIS**.

The auditor reviewed the corrective action plan and accepted it. "This structured approach not only addresses the NC but also positions you as a reliable partner for complex aerospace projects," the auditor remarked.

## Tangible Benefits

The systematic approach to input verification brought immediate and long-term benefits to SkyTech Solutions:

1. **Reduced Rework:** Clear, verified inputs minimized errors and rework, improving overall project efficiency.
2. **Timely Delivery:** By eliminating input-related delays, the company was able to meet project milestones and strengthen customer trust.
3. **Improved Customer Relationships:** The US-based client expressed appreciation for the proactive measures, leading to discussions about potential new projects.
4. **Enhanced Team Confidence:** The new procedures empowered employees to identify and address input-related issues effectively.

## A Year Later

During the follow-up audit, the auditor was impressed by how the changes had been embedded into the organization's processes. The input verification procedures had become a cornerstone of SkyTech's project management framework.

The auditor noted, "Your approach to managing inputs demonstrates a clear understanding of the importance of systematic planning in delivering complex aerospace projects."

SkyTech Solutions not only secured its **AS9100D certification** but also built a reputation as a meticulous and dependable service provider in the global aerospace industry. The experience served as a powerful reminder of the value of starting with clarity and building a solid foundation for success.

## CS 51 [AS9100D Clause 8.3.2] Soaring Through Standards: A Veteran's Quest for Certification

The sun blazed over the sprawling hangar of AeroVista Dynamics, a fledgling private company with ambitious dreams of launching a short-range business jet. Inside, amidst the hum of machines and the rhythmic tapping of keyboards, a man stood at the heart of it all. He was Krishnan Iyer, the Chief Designer—a proud graduate of the Madras Institute of Technology and a veteran of aircraft design from the 1970s. His eyes, weathered by decades of experience, gleamed with the same passion he had carried since his youth.

Krishnan had spent his life shaping the wings of countless aircraft, yet he now faced a challenge unlike any other. Corporate headquarters had issued a daunting directive: achieve AS9100D certification for the design division within three months. For a man rooted in traditional design practices and only vaguely familiar with ISO 9001, the complexities of AS9100D felt like navigating an uncharted sky.

Undeterred, Krishnan and his small team embarked on the Herculean task of preparing for certification. He meticulously aligned their existing processes with his understanding of ISO 9001, believing he had covered the essentials. Yet, the labyrinthine requirements of AS9100D extended far beyond what he had anticipated, especially in the realm of D&D planning.

The stage 2 audit arrived quicker than anticipated. The lead auditor, Richard Caldwell, a seasoned expert from the UK, carried himself with the quiet authority of someone who had seen it all. Richard volunteered to audit the core business process of aircraft design, keen to see how a start-up aimed to break into the competitive aerospace industry.

It did not take long for the cracks to appear.

Richard, with the precision of a surgeon, identified an NC under clause 8.3.2. The D&D process lacked defined activities, clear tasks, resource allocation, and a structured understanding of inputs and outputs. There was no documented evidence that design planning considered verification, testing, and maintenance requirements—crucial elements in the aerospace sector, vide clause 8.3.2.

Krishnan listened in silence as Richard delivered the findings.

"Your team is systematic, and your knowledge is undeniable," Richard said, his voice steady yet empathetic. "But without addressing these gaps, the excellence of your process remains invisible. Fix this, and you do not just earn certification; you set the foundation for sustainable growth."

The words hit Krishnan like a gust of cold wind. But within moments, his resolve hardened. He gathered his team, explaining every nuance of the NC. They dissected clause 8.3.2, studied aerospace-specific requirements, and meticulously restructured their processes. Each design stage was broken into distinct activities with clear documentation of tasks, resources, and responsibilities. The team worked through nights, driven by an almost palpable collective determination.

Weeks later, Richard returned to verify the corrective actions. He pored over documents, questioned team members, and observed processes firsthand. By the end of the day, he walked into Krishnan's office, extended his hand, and offered a rare smile.

"You did it," he said simply. "Certification granted."

As the news spread, the hangar echoed with cheers. Krishnan, standing amid his celebrating team, looked out at the partially assembled jet. They had conquered the mountain of AS9100D, not just for a certificate, but to build something truly enduring.

AeroVista Dynamics had earned its wings, and now, it was ready to soar.

## CS 52 [AS9100D Clause 8.3.3] Transforming Helicopter Design: A Systematic Path to Speed

In Pune, India, RotorWings Aerotech Pvt. Ltd., a burgeoning helicopter design organization, was embarking on an ambitious journey. The company had secured two critical projects: a cutting-edge rotorcraft design for the Indian defense services and a versatile, lightweight helicopter intended for a civilian air taxi service. As the company sought AS9100D certification, its processes were put under scrutiny during the initial certification audit.

### The NC

The audit identified a significant gap under clause 8.3.3:

> *Controls established for design reviews (preliminary review and critical review, subsequent changes and improvements) were not established, except for getting final clearance from the chief of design.*

The auditor noted that while the Chief of Design, Raghav Mehta, had the technical expertise to approve final designs, the absence of structured preliminary and critical design reviews (CDRs) was a risk. This lack of formal controls made it challenging to identify and address design flaws early in the development process, potentially leading to delays, rework, and cost overruns.

### Challenges Faced

The absence of systematic design reviews resulted in:

- **Delayed Issue Identification:** Design flaws and requirement mismatches were often discovered late in the process, causing rework.
- **Inefficient Collaboration:** Teams lacked opportunities for structured input during key stages of the design process.
- **Extended Timelines:** Without clear checkpoints, projects experienced delays, jeopardizing delivery schedules for both defense and civilian customers.

### Immediate Containment

To address the auditor's findings, Mehta took immediate steps:

1. **Temporary Review Panels:** Cross-functional teams were formed to conduct an immediate review of all ongoing designs.
2. **Customer Engagement:** Both defense and civilian clients were briefed on the actions being taken to ensure that their requirements were met without delays.
3. **Documentation Updates:** Existing design documents were updated with any identified gaps or corrective measures.

These actions provided a stopgap measure to address immediate risks while paving the way for long-term improvements.

### RCA

Mehta led a thorough RCA, identifying the following issues:

- **Over-Reliance on Individual Expertise:** The design process heavily depended on the Chief of Design's final approval, with no structured intermediate reviews.
- **Lack of Defined Review Stages:** The absence of clearly defined preliminary and CDR stages led to uncoordinated progress checks.
- **Inadequate Process Awareness:** Design teams were unaware of the importance of systematic reviews in ensuring compliance with AS9100D standards.

### Corrective Actions

With the support of his team and the auditor's guidance, Mehta implemented a series of corrective actions to address the NC:

1. Establishment of Structured Design Review Stages:
    - **Preliminary Design Review (PDR):** Conducted after the initial concept phase to ensure alignment with customer requirements and technical feasibility.
    - **CDR:** Held after detailed designs were prepared, focusing on identifying potential risks and ensuring manufacturability.
2. Creation of Review Panels: Multidisciplinary panels were formed for each project, including representatives from design, quality

assurance, manufacturing, and customer support.

3. Standardized Review Templates: Templates were developed to ensure consistency in capturing review outcomes, action plans, and subsequent follow-ups.
4. Training and Awareness: Teams underwent training on the importance of design reviews, their roles in the process, and how they aligned with AS9100D requirements.
5. Digital Collaboration Tools: Project management software was introduced to track design progress, schedule reviews, and document outcomes, ensuring transparency and accountability.

## Reporting and Auditor Feedback

The corrective actions, supported by evidence such as review records, updated PIFs, and training completion reports, were submitted to **OASIS**. The auditor reviewed and accepted the corrective action plan, commending the company's commitment to improvement.

## Benefits Realized

The systematic approach to design reviews brought significant benefits to RotorWings Aerotech:

1. **Improved Design Quality:** Early identification and resolution of issues reduced rework and enhanced design reliability.
2. **Faster Project Timelines:** Structured reviews streamlined the design process, ensuring timely delivery to both defense and civilian clients.
3. **Enhanced Collaboration:** Cross-functional review panels fostered better communication and teamwork across departments.
4. **Stronger Customer Trust:** The Indian defense services and the civilian customer appreciated the proactive measures, boosting confidence in the company's capabilities.
5. **Operational Efficiency:** The introduction of digital tools improved tracking and documentation, making processes more efficient and transparent.

## A Year Later

During the follow-up audit, the auditor noted significant progress. The design review process was now a model of systematic planning and execution. Projects were advancing smoothly, with both customers praising the company's professionalism and adherence to timelines.

The auditor remarked, "RotorWings Aerotech has transformed its design function, setting a benchmark for systematic progress and efficiency in aerospace design."

For RotorWings Aerotech Pvt. Ltd., the audit findings were a blessing in disguise. The lessons learned not only secured the company's **AS9100D certification** but also positioned it as a trusted partner for high-stakes aerospace projects. The journey from reliance on individual expertise to a robust, collaborative design process became the cornerstone of its success.

## CS 53 [AS9100D Clause 8.3.4] Flight to Excellence: Improving Helicopter Design Reviews

AeroSwift Aviation Pvt. Ltd., a promising helicopter design firm based in Bengaluru, India, had recently embarked on two ambitious projects. The first was a tactical helicopter for the Indian defense services, designed for reconnaissance and rapid deployment. The second was a lightweight, eco-friendly helicopter intended for a civilian air taxi service, an innovative concept to redefine urban transportation.

As the company prepared for its AS9100D certification, its processes underwent a rigorous initial certification audit. A critical NC under clause 8.3.4 was identified:

> *Controls established for design reviews (preliminary review and critical review, subsequent changes and improvements) were not established, except for getting final clearance from the chief of design.*

This finding became a pivotal moment for AeroSwift, spurring transformative changes in its design process.

### The NC

The audit revealed a major procedural gap:

1. Over-Reliance on Final Approval: Design reviews were limited to the Chief of Design, Vivek Gowda, granting final clearance. Intermediate reviews, such as PDR and CDR, were missing.
2. Lack of Change Management: Changes and improvements during the design process were not systematically documented or reviewed.
3. Customer Impact: Both the defense and civilian projects faced potential delays and risks due to the absence of structured review mechanisms.

### Initial Containment

Gowda, a seasoned aeronautical engineer, took immediate containment actions:

1. **Temporary Review Teams:** Multidisciplinary teams were formed to conduct reviews for ongoing projects, ensuring critical gaps were addressed.
2. **Customer Communication:** Both the defense and civilian clients were informed of the measures being taken to ensure quality and timely delivery.
3. **Interim Documentation:** All changes and improvements were documented and reviewed retroactively to create a baseline for further work.

### RCA

The design team, led by Gowda, conducted a comprehensive RCA and identified the following:

- Process Deficiency: The lack of formal design review stages was a systemic issue rooted in an informal, ad-hoc approach.
- Time Constraints: The team focused on meeting delivery timelines, often skipping intermediate reviews to expedite the process.
- Limited Awareness: The importance of structured design reviews under AS9100D was not fully understood across the organization.

### Corrective Actions

With insights from the RCA, the team developed a robust corrective action plan:

1. Introduction of Structured Design Reviews:
   - **PDR:** Conducted after initial design concepts were developed to ensure alignment with customer requirements and feasibility.
   - **CDR:** Focused on detailed designs, manufacturability, and risk mitigation before finalizing the design.
2. Change Management Process: A formal process for documenting and reviewing design changes was implemented, ensuring traceability and accountability.
3. Multidisciplinary Review Panels: Panels included representatives from design, quality assurance, manufacturing, and customer support, ensuring diverse perspectives during reviews.
4. Standardized Templates and Tools: Templates for recording review outcomes and action plans were introduced, supported by project management software to track progress.
5. Training and Awareness: Training sessions were conducted to educate teams about AS9100D requirements and the importance of systematic design reviews.

### Reporting and Auditor Feedback

The corrective actions, along with supporting evidence such as updated PIFs, review records, and training logs, were submitted to **OASIS**. The auditor reviewed the plan and evidence, commending AeroSwift's commitment to improvement.

### Benefits Realized

The changes brought significant benefits to AeroSwift Aviation:

1. **Enhanced Design Quality:** Issues were identified and resolved early in the process, reducing rework and ensuring robust designs.
2. **Streamlined Progress:** Structured reviews provided clear milestones, improving efficiency and reducing project timelines.
3. **Customer Confidence:** Both defense and civilian customers expressed satisfaction with the improved processes, strengthening their trust in AeroSwift's capabilities.
4. **Operational Efficiency:** The use of digital tools and standardized templates improved collaboration and documentation across teams.

### A Year Later

During the follow-up audit, the auditor noted significant improvements in AeroSwift's design processes. The structured design reviews had become an integral part of the organization's culture, ensuring systematic progress and quality in all projects.

The auditor remarked, "Your approach to implementing design controls demonstrates a strong commitment to delivering excellence, for both defense and civilian aerospace applications."

For AeroSwift Aviation, the NC turned into an opportunity for growth. The lessons learned not only secured their **AS9100D certification** but also positioned them as a trusted leader in helicopter design. The systematic approach laid a foundation for future success, enabling the company to soar to new heights in the aerospace industry.

## CS 54 [AS9100D Clause 8.3.4.1] New Flight Path: Revamping Design Validation at AeroTech

AeroTech Designs Pvt. Ltd., a leading aircraft design organization based in Hyderabad, India, was entrusted with a high-stakes project: designing a state-of-the-art reconnaissance aircraft for the Indian defense services. Known for their technical expertise, the team had a reputation for innovation but faced a crucial test during their AS9100D certification audit.

### The NC

The audit uncovered a significant NC under **clause 8.3.4.1 of AS9100D**, which read:

1. **Consolidation of Test Plans and Results:** The auditor found no systematic approach to consolidating test plans, results, and acceptance criteria, making verification challenging.
2. **Final Configuration Confirmation:** There was no assurance that the items tested were in their final, approved configuration.
3. **Calibration of Measuring Devices:** The absence of documented confirmation that calibrated measuring devices were used to record test results was a critical gap.

These findings highlighted systemic inefficiencies in AeroTech's design validation processes, posing risks to both the project timeline and product quality.

### Initial Containment

Dr. Arvind Kumar, the Chief of Design, took immediate action:

1. **Freezing Configurations:** A configuration control protocol was introduced to ensure only the final approved designs were subjected to testing.
2. **Equipment Audit:** All measuring devices were audited for calibration status, and uncalibrated devices were flagged for immediate recalibration or replacement.
3. **Customer Communication:** The Indian defense services were informed of the proactive measures taken to ensure compliance and mitigate any potential risks.

### RCA

A cross-functional team conducted a thorough RCA and identified:

1. **Fragmented Documentation:** Test plans, results, and acceptance criteria were spread across multiple files, leading to inefficiencies in verification.
2. **Informal Configuration Control:** The lack of a formal mechanism to verify the final configuration of test items caused inconsistencies.
3. **Lapses in Calibration Oversight:** Inadequate tracking of calibration schedules for measuring devices contributed to noncompliance.

### Corrective Actions

To address the root causes, AeroTech implemented a comprehensive corrective action plan:

1. Centralized Design Validation Repository:
    - All test plans, results, and acceptance criteria were consolidated into a centralized digital repository, accessible to all relevant stakeholders.

- A standardized format was developed to ensure consistency and clarity in documentation.

2. Configuration Control Protocol:
   - A formal procedure was introduced to verify and document the final configuration of items before testing.
   - Configuration audits were integrated into the design process to ensure compliance.
3. Enhanced Calibration Process:
   - A calibration management system was implemented to track the calibration status of all measuring devices.
   - Only calibrated devices were authorized for use in design validation, with calibration certificates attached to test reports.
4. Training and Awareness:
   - Training sessions were conducted to educate the design and testing teams on AS9100D requirements, emphasizing the importance of systematic validation and calibration.

## Reporting and Auditor Feedback

The corrective actions, along with supporting evidence such as consolidated test documentation, configuration control records, and calibration certificates, were submitted to **OASIS**. The auditor reviewed the plan and evidence, commending AeroTech's systematic approach to addressing the NC.

## Benefits Realized

The changes brought transformative benefits to AeroTech's operations:

- **Streamlined Validation Process:** The centralized repository reduced duplication and errors, improving efficiency and accuracy.
- **Improved Design Integrity:** Configuration control ensured that tests reflected the final design, boosting confidence in the results.
- **Compliance Assurance:** The calibration management system eliminated risks associated with uncalibrated devices, ensuring compliance with AS9100D.
- **Customer Confidence:** The Indian defense services appreciated AeroTech's commitment to quality and systematic improvements, strengthening their partnership.

## A Year Later

During the follow-up audit, the auditor observed that AeroTech had fully integrated the corrective actions into their processes. The centralized validation system and configuration controls had become benchmarks for operational excellence.

The auditor noted, "AeroTech Designs has not only resolved the NC but has set a new standard in design validation, ensuring both quality and efficiency in delivering mission-critical aircraft to the Indian defense services."

For AeroTech Designs, the NC was a turning point. By addressing systemic gaps, the company not only secured its **AS9100D certification** but also established a robust foundation for future projects. The lessons learned positioned AeroTech as a leader in aircraft design, ready to meet the evolving needs of the aerospace industry.

## CS 55 [AS9100D Clause 8.3.5] Building Precision: Closing Design Gaps at Skyline Aerotech

Skyline Aerotech Pvt. Ltd., an aircraft design organization based in Bengaluru, India, was in the spotlight as it embarked on an ambitious project to design an advanced surveillance aircraft for the Indian defense services. The initial certification audit for AS9100D exposed a critical flaw in their design processes, bringing their pursuit of excellence to a pivotal moment.

### The NC

The audit uncovered a major NC under clause 8.3.5 of AS9100D, revealing:

1. The gap in critical items identification in drawings/specifications.
2. The design outputs failed to specify critical items, including key characteristics and specific actions to be taken for these items in the outputs.
3. Lack of an approval process for the release of the outputs.

Design outputs were not consistently approved by authorized personnel before release, raising concerns about the accountability and integrity of the process.

The auditor emphasized that these gaps could lead to serious consequences, including potential safety risks and compromised project deliverables if left unaddressed.

### Immediate Containment

Dr. Neeraj Sinha, the Chief of Design at Skyline Aerotech, took swift action:

- Temporary Hold on Design Releases: All ongoing design releases were paused to ensure compliance with AS9100D requirements.
- Critical Item Review: An immediate review of existing designs was initiated to identify and document critical items and their key characteristics.
- Approval Protocols: A temporary manual process was implemented to ensure all design outputs were reviewed and approved by authorized personnel before release.

### RCA

The Chief of Design led a cross-functional team in a detailed RCA, which revealed:

- **Lack of a Defined Framework:** There was no structured approach to identifying critical items or their associated key characteristics in design outputs.
- **Informal Approval Practices**: The absence of a formalized approval workflow led to inconsistencies in the review and release process.
- **Inadequate Training:** Design teams were unaware of the specific requirements outlined in clause 8.3.5, leading to unintentional noncompliance.

### Corrective Actions

Skyline Aerotech developed and implemented a robust corrective action plan:

1. Critical Item Identification Process:
   - A structured framework was introduced to identify critical items in all design outputs, with detailed documentation of key characteristics and associated actions.
   - A checklist was developed to ensure no critical items were overlooked during the design process.

2. Approval Workflow Automation:
   - A digital approval system was implemented to streamline the review process, ensuring all design outputs were approved by authorized personnel before release.
   - The system included automated alerts and reminders to prevent delays.
3. Team Training:
   - Comprehensive training sessions were conducted to educate the design teams on AS9100D requirements, with a focus on clause 8.3.5.
   - The training emphasized the importance of critical item identification and proper documentation.
4. Design Review Integration:
   - The design review process was enhanced to include checkpoints for critical items and approval status, ensuring compliance at every stage.

## Reporting and Auditor Feedback

The corrective actions, along with supporting evidence such as updated design documentation, approval records, and training logs, were submitted through OASIS.

The auditor reviewed the plan and supporting documents, noting the significant improvements. They highlighted the proactive approach Skyline Aerotech had taken to address the NC and improve their processes.

## Long-Term Benefits

The systematic changes brought transformative benefits to Skyline Aerotech:

- **Enhanced Design Integrity:** The structured approach to identifying and managing critical items ensured robust and reliable designs, meeting the exacting standards of the Armed Forces.
- **Streamlined Approvals:** The automated approval system reduced delays and improved accountability, fostering a culture of precision and efficiency.
- **Increased Customer Confidence:** The visible commitment to quality and compliance reassured the Indian defense services of Skyline Aerotech's capabilities and reliability.
- **Team Empowerment:** Training sessions not only ensured compliance but also enhanced the technical expertise of the design teams, boosting morale and productivity.

## A Year Later

During the follow-up audit, the auditor observed that the corrective actions had been seamlessly integrated into Skyline Aerotech's processes. The team's commitment to quality and continual improvement was evident, earning the organization accolades and securing their AS9100D certification.

The auditor concluded, "Skyline Aerotech's approach to addressing the NC demonstrates their dedication to excellence. The improvements made not only ensure compliance but also position the organization as a leader in aircraft design."

For Skyline Aerotech, the NC was a turning point that spurred a transformation in its design processes, reinforcing its reputation as a trusted partner in the aerospace industry.

## CS 56 [AS9100D Clause 8.3.6] Strengthening Configuration Management: AeroSwift's Turning Point

AeroSwift Innovations Pvt. Ltd., a leading aircraft design organization based in Bengaluru, India, was entrusted with an ambitious project to design advanced surveillance aircraft for the Indian defense services. As part of their pursuit of AS9100D certification, the organization underwent an initial certification audit, which revealed a critical NC under clause 8.3.6.

### The NC

The audit highlighted deficiencies in the organization's **configuration management** process:

1. **Ineffective Design Change Management:** Design changes were not effectively managed, leading to discrepancies in part drawings issued for production.
2. **Outdated Production Drawings:** Production teams were working with outdated versions of drawings that did not reflect the latest design changes, increasing the risk of manufacturing errors.

The auditor emphasized that these gaps could jeopardize product quality and safety, compromising the organization's ability to meet customer requirements and maintain compliance.

### Immediate Containment

The Chief of Design, Rajiv Menon, took decisive action:

1. **Temporary Production Halt:** All production activities were temporarily paused to prevent the manufacture of parts based on outdated drawings.
2. **Drawing Version Review:** An immediate review was conducted to identify discrepancies between design revisions and production drawings.
3. **Communication to Teams:** The design and production teams were notified about the issue, and updated drawings were urgently circulated.

### RCA

A cross-functional team was formed to investigate the root cause of the issue. The analysis revealed:

- **Lack of Real-Time Updates:** The system for managing design changes was not integrated with production, causing delays in updating drawings.
- **Manual Processes:** The reliance on manual processes for tracking design changes led to errors and oversight.
- **Inadequate Training:** Teams were not adequately trained in configuration management practices, leading to inconsistencies in maintaining version control.

### Corrective Actions

AeroSwift Innovations implemented a comprehensive corrective action plan:

1. Centralized CMS:
   - A digital **CMS** was introduced to automate the tracking and management of design changes.
   - The CMS ensured that production teams had access to the latest approved drawings in real time.
2. Version Control Protocols:
   - A robust version control system was implemented to ensure that only the latest design versions were issued for production.

- Each design change was accompanied by a mandatory verification and approval process.

3. Training Programs:
   - Training sessions were conducted for design and production teams to enhance their understanding of configuration management and the importance of version control.
4. Design–Production Integration:
   - The design and production teams were integrated into a unified workflow to improve communication and coordination.
   - Regular cross-functional reviews were introduced to ensure alignment.

### Reporting and Auditor Feedback

The corrective actions, along with supporting evidence such as updated configuration management protocols, training records, and CMS implementation reports, were submitted through **OASIS**.

The auditor reviewed the plan and acknowledged the organization's proactive approach to addressing the NC. The improvements were noted as a positive step toward ensuring compliance and enhancing operational efficiency.

### Long-Term Benefits

The changes brought about by the audit findings resulted in significant improvements for AeroSwift Innovations:

- Error-Free Production: The CMS ensured that only the latest, approved designs were used in production, reducing errors and rework.
- Enhanced Efficiency: Automation of configuration management streamlined processes, saving time and resources.
- Increased Reliability: The improved alignment between design and production teams enhanced the reliability of the organization's outputs, building trust with the Indian defense services.
- Stronger Compliance Culture: The training programs fostered a culture of compliance and continual improvement within the organization.

### A Year Later

During the follow-up audit, the auditor noted the seamless integration of the corrective actions into AeroSwift's processes. The organization was commended for its commitment to quality and its ability to adapt to challenges. The auditor remarked, "AeroSwift Innovations has turned a critical NC into an opportunity for transformation. The enhanced CMS is a benchmark for other organizations."

This experience reinforced AeroSwift's reputation as a trusted partner in the aerospace industry, securing their certification and setting a solid foundation for future projects.

## CS 57 [AS9100D Clause 8.4.1] Learning Valuable Lessons in Supplier Risk Management

A Hyderabad-based software company was undergoing its certification audit for the AS9100D. This marked a significant milestone as they entered the aerospace business, a sector renowned for its stringent quality and process requirements. The firm's Purchase Manager Sampath Kumaran had implemented a system for managing externally provided processes, products, and services. However, during the audit, the auditor uncovered a critical gap.

### NC Issued

*The auditor issued an NC citing clause 8.4.1 of AS9100:*

*"The organization has not identified and managed the risks associated with the suppliers of processes, products, and services, as well as the selection and use of external providers."*

This shortfall exposed the organization to potential risks such as delays, quality issues, or noncompliance with aerospace standards—risks that could jeopardize customer satisfaction and timely completion of job orders.

### Immediate Response and Proposed Actions

Recognizing the gravity of the issue, the organization committed to addressing the NC promptly. The Purchase Manager, along with the quality team, proposed the following actions:

1. Risk Assessment of Existing Suppliers:
    - Conduct an immediate evaluation of all current suppliers to identify risks in areas such as quality, delivery timelines, and compliance with aerospace requirements.
    - Document the risks and their potential impacts on ongoing projects.
2. RCA:
    - Investigate why risks associated with suppliers were not previously identified or managed.
    - Review the organization's supplier selection, approval, and monitoring processes for gaps.
3. Corrective Actions:
    - Define clear criteria for evaluating and selecting suppliers, focusing on their capability to meet aerospace standards.
    - Implement a robust risk management process to regularly monitor and mitigate supplier risks.
    - Establish contracts with suppliers that include clauses for risk mitigation, performance guarantees, and penalties for noncompliance.
4. Evidence Submission:
    - Document and submit evidence of the actions taken to the auditor within the timeframe demanded by the OASIS.

### Auditor's Review

After some time, the auditor revisited the organization to review the corrective actions. The following observations were made:

1. Risk Assessment of Suppliers:
   - The organization had completed a thorough risk assessment of all existing suppliers.
   - High-risk suppliers were flagged, and mitigation plans were documented, such as requiring additional quality checks or transitioning to more reliable suppliers.
2. RCA and Corrective Actions:
   - The RCA identified a lack of structured supplier evaluation and risk management processes as the primary issue.
   - The organization implemented a supplier evaluation and approval system, including risk assessment as a mandatory step during selection.
3. Implementation and Monitoring:
   - Evidence of implementation, such as risk registers, updated supplier performance reviews, and new supplier contracts, was provided.
   - Training sessions were conducted for the purchasing and quality teams to ensure consistent application of the new processes.

Satisfied with the actions taken, the auditor accepted the corrective actions and recorded them in the audit report. However, the auditor noted that the **effectiveness of these measures** would be verified during the next surveillance audit.

### Lessons Learned

This experience highlighted critical lessons for the organization:

- **Importance of Supplier Risk Management:** In the aerospace sector, the stakes are high, and even minor supplier issues can lead to delays, cost overruns, or reputational damage. A proactive approach to risk identification and management is crucial.
- **Structured Supplier Evaluation and Monitoring:** Selecting suppliers based on documented criteria and regularly monitoring their performance ensures consistent quality and reliability.
- **Timely and Comprehensive Corrective Actions:** Addressing NCs promptly with a systematic approach not only ensures compliance but also strengthens the organization's overall processes.
- **Continual Improvement:** The organization realized the need for a culture of continual improvement, where lessons learned from audits are integrated into everyday operations.

### Looking Ahead

With improved systems in place, the organization was better positioned to meet aerospace requirements and customer expectations. The management recognized the value of rigorous processes in achieving long-term success in the aerospace sector. As the surveillance audit loomed in the coming year, the company was confident in its ability to demonstrate the effectiveness of its corrective actions and build stronger relationships with its foreign customers.

This journey not only paved the way for certification but also instilled a deeper understanding of the role of supplier risk management in ensuring operational excellence and customer satisfaction.

## CS 58 [AS9100D Clause 8.4.1] When Trust Got in the Way of Quality

### Scene 1: The Inspection Floor

The clock ticked loudly in SHOHUM's receiving inspection area as Ramesh, the company's senior inspector, scanned a freshly delivered crate of components. The items bore the stamp of a prestigious customer—an emblem of trust that SHOHUM prided itself on.

Ramesh turned to his assistant and grinned. "Straight from the customer! No need for checks. Let's label these and send them to storage."

The assistant hesitated. "But... should not we at least look at the specs?"

Ramesh laughed. "Listen, the customer *knows* what they're doing. If they're sending it, it is good to go. Trust me, kid. I've been doing this for years."

Unbeknownst to them, a silent storm was brewing.

### Scene 2: The Audit Begins

Three weeks later, SHOHUM was abuzz with activity. Their AS9100D certification audit had begun. Rajesh, a meticulous and sharp-eyed auditor from India, was poring over their documentation.

Rajesh paused as he reached the receiving inspection procedure. He raised an eyebrow and called for SHOHUM's Quality Manager, Aisha.

"This procedure..." Rajesh began, his tone calm but firm, "...says you do not verify customer-supplied products?"

Aisha nodded confidently. "That's correct. We trust our customers to supply exactly what's needed for their equipment. They're experts in their domain."

Rajesh set the document down and leaned forward. "Aisha, do you understand what clause 8.4.1 of AS9100D requires?"

Aisha blinked. "Of course. We're responsible for the conformity of all externally provided products."

"Exactly," Rajesh said, his voice tightening. "Including those defined by your customers. Tell me—how do you ensure that these components meet specifications?"

"Well, we... do not," Aisha admitted, her confidence wavering. "But they're the customer! Why would they send something that doesn't work?"

Rajesh frowned. "Let me put it this way. If a customer-supplied component fails and causes an issue in your product, who do you think will bear the blame?"

Aisha's face paled. "We would."

"Exactly. Trust is important, but so is verification. And right now, you're leaving the door wide open for risk."

### Scene 3: The Realization

Later that afternoon, Rajesh gathered the SHOHUM team in the conference room. The atmosphere was tense.

Rajesh stood at the head of the table. "Let me share a story with you."

The team leaned in, curious.

"Years ago, I audited a company in India. They had a similar policy—accepting customer-supplied products without verification. One day, a customer-provided component failed, causing a catastrophic issue. Their entire production line was halted, they lost millions in revenue, and their reputation took a hit. Why? Because they assumed the customer's product was flawless. Assumptions are dangerous."

Ramesh, sitting at the back, shifted uncomfortably. "But sir, customers know what they want, do not they?"

Rajesh's gaze sharpened. "Do you think customers never make mistakes? Or that their suppliers are infallible? The standard exists for a reason—to prevent failures before they happen."

Aisha spoke up, her voice steady but tinged with regret. "What do we need to do to fix this?"

Rajesh nodded, pleased. "Good question. First, you need to recognize this as a major NC. Then, you'll need to revise your procedures, perform RCA, and implement corrective actions."

### Scene 4: Closing the Gap

Over the next three weeks, SHOHUM worked tirelessly.

- Aisha spearheaded the revision of the receiving inspection procedure. "From now on," she declared in a team meeting, "every product, regardless of its source, will be inspected and verified. No exceptions."
- Ramesh organized immediate training sessions for the inspection team. He stood before them, humbled. "I was wrong to assume customer-supplied means perfect. We're the last line of defense for quality."
- The team conducted a thorough quality check of all customer-supplied components in storage. Two components were found to have minor deviations from specifications—a sobering reminder of their past oversight.

When the corrective actions were completed, Aisha uploaded the evidence to the IAQG/OASIS portal.

### Scene 5: The Follow-Up

Three weeks later, Rajesh reviewed the uploaded documentation and returned for a meeting with the team.

"You've done well," he said, a rare smile crossing his face. "You've closed the gap and demonstrated your commitment to quality. But remember, this is not just about passing an audit. It is about building a culture where quality comes first—always."

Aisha nodded. "We've learned our lesson, Rajesh. Trust is important, but verification is non-negotiable."

Rajesh's eyes sparkled with approval. "Good. Because next year, during your surveillance audit, I'll be checking not just your procedure but its effectiveness. Keep up the good work."

As Rajesh departed, the SHOHUM team exchanged glances, a newfound determination in their expressions. They had learned a hard but invaluable lesson: quality is not just a standard to meet—it is a principle to live by.

### The Moral of the Story

This tale of oversight and recovery serves as a reminder: trust your customers but verify everything. In the high-stakes world of manufacturing, assumptions can lead to disaster, but diligence ensures success.

## CS 59 [AS9100D Clause 8.4.1] Real-Life Illustration: The Hydraulic Pump Mishap

In the high-stakes world of aerospace manufacturing, quality and compliance are paramount. However, there are times when operational pressures can cloud judgment, leading to critical oversights. This is a story of such an oversight and the lessons learned to prevent it from recurring.

### The Context

A prestigious customer had ordered a new batch of trainer aircraft to meet expanding operational needs. The original hydraulic pump used in these aircraft had gone out of production. The customer, keen on expediting the delivery, identified an alternative pump and urged the manufacturer to adopt it. Under significant operational pressure, the manufacturer agreed to the substitution without thoroughly assessing the pump's suitability for its intended application.

### The Oversights

The alternative pump was procured hastily—34 units in total. No comparative study or testing was conducted to verify whether the new pump could meet the environmental and operational demands specific to the tropical conditions where the aircraft would operate. Despite the manufacturer's awareness of best practices outlined in **AS9100**, these were overlooked due to the urgency of the project.

Key lapses included:

1. Failure to Evaluate Suitability: The substituted pump's compatibility with the aircraft and its operational environment was not validated.
2. No Inspection or Testing: The manufacturing team did not conduct basic inspections to verify that the pump met the required specifications or matched the original design's connectors.
3. Modifications Without Validation: When the connectors were found to be incompatible, local modifications were made without proper validation or certification.

### The Fallout

During the first test flight of the newly assembled trainer aircraft with the alternative pump, the hydraulic power system failed. This failure mirrored similar incidents documented in the French Airworthiness Directive No. F-2007-010 (dated September 12, 2007), highlighting the risks of unverified substitutions.

The test flight failure underscored the gravity of the oversight, leading to delays, additional costs, and reputational risks.

### The Certification Audit

Shortly after the incident, a scheduled certification audit under AS9100D exposed the gaps in the manufacturer's QMS. Clause **8.4.1** was identified as a major NC.

The auditor cited:

- The lack of proper evaluation of the substituted pump.
- Inadequate controls over procurement and quality assurance.
- A systemic failure to adhere to the organization's own QMS standards.

(See note in AS9100D below clause 8.4.2. Customer verification activities performed at any level of the supply chain do not absolve the organization of its responsibility to provide acceptable processes, products, and services and to comply with all requirements.)

He also issued a NC vide clause 8.4.1 "The organization shall be responsible for the conformity of all externally provided processes, products, and services, including from sources defined by the customer."

## Corrective Actions

To address the NC and align with **clause 10.2.1** (NC and corrective action), the following steps were implemented:

**1.** RCA:
   - Identified that operational urgency and lack of a risk-based approach led to the oversight.
   - Recognized insufficient training and awareness of QMS practices among procurement and engineering teams.

**2.** Revised Procurement Procedures:
   - Introduced mandatory risk assessments for all substituted parts.
   - Enforced comparative studies, including environmental testing, before procurement.

**3.** Training and Awareness:
   - Conducted extensive training sessions on AS9100 requirements for all stakeholders.
   - Emphasized the importance of adhering to QMS principles even under operational pressure.

**4.** Improved Quality Checks:
   - Established stringent inspection protocols for all incoming components.
   - Mandated compatibility and functional testing before integration into aircraft systems.

**5.** Enhanced Communication:
   - Improved coordination with customers to ensure clarity on quality expectations.
   - Instituted joint reviews for any customer-recommended changes.

## The Outcome

The corrective actions were reviewed and accepted by the auditor. Their effectiveness was scheduled for verification during the next surveillance audit. The incident served as a turning point, reinforcing the importance of QMS adherence and risk management in aerospace manufacturing.

## The Takeaway

Quality cannot be compromised, especially in aerospace. This incident serves as a reminder that "shortcuts" can lead to significant risks, including financial loss, operational delays, and, most importantly, safety hazards. By implementing a robust QMS aligned with AS9100D, manufacturers can ensure that quality and compliance are maintained, even under challenging circumstances. The organization shall verify the quality of all externally provided processes, products, and services, including those from sources defined by the customer or supplied by the customer before acceptance.

## A Lesson Learned

This case exemplifies the need for diligence, even when operational demands are high. By fostering a culture of quality, organizations can prevent such costly errors, safeguarding both their reputation and their customers' trust.

## CS 60 [AS9100D Clause 8.4.1.1] Fixing NC in Supplier Evaluation and Purchasing

In an SME in Japan, the raw material purchasing process had been managed with an exceptional level of trust and minimal documentation. The MD relied on a trade directory to identify suppliers, communicated specifications over the phone, and promptly paid electronically after receiving the materials. The system functioned smoothly due to the inherent honesty of the parties involved. However, during an audit for certification to AS9100, an Indian auditor identified critical gaps in the purchasing process.

The auditor noted that the organization lacked a formal mechanism to evaluate and document supplier capabilities, maintain an approved suppliers list, or monitor supplier performance. These deficiencies conflicted with **clause 8.4.1.1 of AS9100**, which emphasizes structured supplier management processes. Consequently, the auditor issued an **NC** highlighting the absence of:

1. A register of external providers, including approval status and scope.
2. Supplier performance reviews on quality and OTD.

### NC Details

*The NC stated: "The organization has not:*

*(a) Maintained a register of its external providers that includes approval status (e.g., approved, conditional, disapproved) and the scope of the approval (e.g., product type, process family).*

*(b) Reviewed the suppliers' performance on quality and OTD."*

The auditor explained the criticality of these requirements in ensuring that suppliers consistently meet specified quality and delivery standards. While appreciating the organization's operational efficiency and trust-based approach, the auditor emphasized the need to align processes with the formal requirements of the standard.

### Organization's Response and Follow-Up Actions

The MD, eager to achieve certification, acknowledged the findings and took the following corrective actions:

1. Establishment of a Supplier Register:
    - Created a centralized supplier register with details such as supplier names, approval status (approved, conditional, disapproved), and scope of approval (e.g., raw materials, machining processes).
    - Included historical performance data and documentation of initial evaluations.

2. Implementation of a Supplier Evaluation System:
   - Developed criteria to evaluate and approve suppliers based on their ability to meet specified requirements.
   - Conducted a review of all existing suppliers to assess their capabilities and documented the findings in the supplier register.
3. Supplier Performance Monitoring:
   - Implemented a system to regularly monitor supplier performance on key parameters such as quality and OTD.
   - Introduced a scoring mechanism to classify suppliers into categories, triggering actions such as re-evaluation or supplier development programs.
4. Formalization of the Purchase Process:
   - Revised the purchase process to include issuing formal requests for quotations (RFQs) to approved suppliers, vetting offers, and issuing detailed POs with adequate information.
   - Integrated verification of purchased materials and conformity certificates upon receipt.
5. Training and Awareness:
   - Conducted training sessions for purchasing staff to ensure understanding and compliance with the new procedures.
   - Engaged the auditor to provide guidance on establishing a robust supplier management framework.

### Auditor's Follow-Up and Verification

During the subsequent visit, the auditor reviewed the corrective actions and verified the following:

1. The supplier register was comprehensive and up to date, with clear documentation of approval statuses and scopes.
2. Performance reviews had been conducted, with evidence of corrective actions taken for underperforming suppliers.
3. POs issued post-audit adhered to the revised process, including detailed specifications and traceable approvals.
4. Documentation showed that materials received were checked for quality and conformity before acceptance.

The auditor noted that while the corrective actions were effective on paper, their implementation and sustainability would be monitored during the surveillance audit.

### Conclusion

The organization successfully addressed the NC by implementing robust supplier management and purchasing processes aligned with AS9100. This experience underscored the importance of balancing trust with structured systems, ensuring long-term reliability and compliance in critical industries like aerospace.

## CS 61 [AS9100D Clause 8.4.1.1] The Supplier Performance Oversight

In Singapore's bustling aerospace manufacturing hub, AeroCraft Precision Engineering Pvt. Ltd., a built-to-print organization, was preparing for a routine surveillance audit for AS9100D compliance. Known for its high-quality components supplied to major aerospace clients, the company's operations were under scrutiny to ensure continued adherence to the stringent standards of the aerospace industry.

The lead auditor, Ethan Koh, arrived with a detailed checklist, ready to dive into the company's processes. Among the key areas he intended to examine was supplier management—a critical component under AS9100D clause 8.4.1.1.

### Uncovering the Gap

During his review of the procurement process, Ethan asked to see the records of supplier evaluations, periodic performance monitoring, and the criteria used to issue POs. The Procurement Manager, Rachel Tan, provided a stack of POs and some initial supplier selection documents.

However, when Ethan asked about performance monitoring, Rachel hesitated. "We evaluate suppliers during their initial onboarding, but we do not have a formal system for monitoring their performance over time," she admitted.

Ethan probed further. "How do you ensure suppliers meet quality, delivery, and other requirements consistently? Are POs issued based on their past performance?"

Rachel shook her head. "We do not systematically track or benchmark performance. If there's an issue, we address it informally."

Ethan frowned. "This is a NC under AS9100D clause 8.4.1.1. Supplier performance must be monitored and benchmarked, and records of these evaluations must be retained. Without this, there's no assurance that suppliers are meeting the standards required for aerospace products. Issuing POs without this oversight increases the risk of nonconforming products or delays."

### Issuing the NC and Recommendations

*Ethan issued an **NCR** with the following details:*

***NC clause:** AS9100D clause 8.4.1.1*

***Observation:** Supplier performance is not monitored for benchmarked performance, and POs are issued without considering the performance of previous supplies.*

***Impact:** This lack of monitoring creates risks for nonconforming products, delayed deliveries, and a loss of traceability, which are critical in the aerospace sector.*

He recommended the following actions:

1. **Implement a Supplier Performance Monitoring System:** Develop a structured process for monitoring supplier performance, including quality, delivery adherence, and responsiveness.
2. **Integrate Performance Metrics into Procurement:** Establish benchmarks for supplier performance and use them as criteria for issuing future POs.
3. **Maintain Comprehensive Records:** Retain records of supplier evaluations, periodic performance reviews, and decisions made based on these assessments.
4. **Communicate Performance Expectations to Suppliers:** Share performance benchmarks

and provide feedback to suppliers to encourage continual improvement.

### Corrective Actions and Follow-Up

AeroCraft's leadership responded promptly to the NC, implementing several corrective measures:

- A supplier scorecard system was introduced, evaluating quality, OTD, and adherence to specifications.
- Retrospective evaluations were conducted for existing suppliers, and the results were documented to establish performance baselines.
- POs were linked to supplier performance data, ensuring that only reliable suppliers were contracted for critical components.
- Supplier contracts were updated to include performance requirements and penalties for noncompliance.

### Verification and Closure

When Ethan returned for the follow-up audit, Rachel presented detailed records of supplier evaluations, periodic performance reviews, and the new supplier scorecard system. Ethan also reviewed a sample of recent POs to confirm they were issued based on the documented performance benchmarks.

"This is a significant improvement," Ethan remarked. "You've demonstrated a clear commitment to meeting AS9100D requirements and strengthening your supply chain."

He closed the NC, commending AeroCraft for their proactive approach. The updated supplier management process not only ensured compliance but also enhanced the company's reputation for reliability and quality in the competitive aerospace industry.

## CS 62 [AS9100D Clause 8.4.2] Hyderabad Software Firm's Path to Supplier Excellence

A Hyderabad-based software company, venturing into the aerospace sector, was undergoing its certification audit for AS9100D. This certification was crucial as the company aimed to cater to international aerospace clients, a domain known for its uncompromising quality standards.

The firm's Purchase Manager, Sampath Kumaran, had established a system for managing externally provided processes, products, and services. However, during the audit, the auditor identified a significant oversight in the organization's purchasing process.

### NC Identified

*The auditor issued an **NC** under **clause 8.4.2 of AS9100D**, stating:*

*"Supplier performance is not monitored for benchmarked performance, and POs are issued without monitoring the performance of previous supplies by the supplier."*

This gap posed a risk to the organization's ability to maintain consistent quality, reliability, and timely delivery of projects, all of which are critical in the aerospace sector.

### Follow-Up Actions

In response to the NC, the company implemented corrective actions as outlined in **clause 10.2.1 of AS9100D**. These actions included:

**1.** Immediate Containment Actions:

- Conducted a review of all recent supplier transactions to identify any quality or delivery issues.

- Temporarily halted issuing new POs until supplier performance data were evaluated.

2. RCA:
   - The team identified the root cause as the absence of a formal process to monitor and benchmark supplier performance.
   - Lack of awareness and training on the importance of supplier performance monitoring was also noted.
3. Corrective Actions:
   - Developed a supplier performance monitoring system, including benchmarks for quality, OTD, and responsiveness.
   - Implemented a scoring mechanism to rate suppliers based on past performance.
   - Updated the purchasing process to ensure POs are issued only to suppliers meeting performance criteria.
   - Established a feedback loop to communicate performance data to suppliers, encouraging continual improvement.
4. Training and Awareness:
   - Conducted training sessions for the purchasing and quality assurance teams to ensure they understood the new procedures and benchmarks.
5. Documentation and Evidence Submission:
   - Submitted evidence of the new system, including supplier performance reports and updated purchasing procedures, to the auditor for review.

## Auditor's Review

The auditor reviewed the corrective actions during a follow-up visit and noted the following:

- **Supplier Performance Monitoring System:** A formal system was in place, with performance metrics documented and regularly reviewed.
- **Updated Purchasing Process:** POs were now issued based on supplier performance data, ensuring only reliable suppliers were engaged.
- **Evidence of Implementation:** Supplier scorecards, meeting minutes, and updated procedures were provided as evidence of the corrective actions.

The auditor accepted the corrective actions, commending the organization for addressing the NC comprehensively. However, the auditor emphasized that the **effectiveness of these actions** would be verified during the next surveillance audit.

## Lessons Learned

This experience imparted several valuable lessons to the organization:

- **Importance of Supplier Performance Monitoring:** Regular monitoring of supplier performance is essential to ensure consistent quality, reliability, and adherence to delivery schedules.
- **Data-Driven Decision-Making:** Using performance metrics to guide supplier selection fosters accountability and supports continual improvement.

- **Systematic Approach to Corrective Actions:** Addressing NCs with a structured approach not only resolves the issue but also strengthens the organization's processes.
- **Supplier Collaboration:** Sharing performance feedback with suppliers fosters transparency and encourages them to align with organizational expectations.

### Moving Forward

The software company emerged stronger from the certification audit. With a robust supplier performance monitoring system in place, the organization was better equipped to meet aerospace standards and deliver on customer expectations.

This journey underscored the criticality of aligning supplier performance with organizational goals, ensuring POs are issued to suppliers who consistently meet the company's benchmarks. As the next surveillance audit approached, the company was confident in its ability to demonstrate the effectiveness of its improvements and solidify its position in the aerospace sector.

## CS 63 [AS9100D Clause 8.4.2] Rising to the Challenge

### Prologue: A Vision of Excellence

The dawn of a new chapter unfolded for SkyReach Aviation, an aircraft manufacturing company in China, as it embarked on its first surveillance audit to retain its AS9100 certification. The stakes were high: ensuring a confirming product, zero customer complaints, and continued certification. This was the foundation of SkyReach's credibility in the global aerospace sector.

The audit team from a German CB, led by Dr. Klaus Reinhardt, an experienced aerospace auditor, arrived with the intent to examine every process thoroughly. Among them was Rajesh Varma, an Indian auditor known for his attention to detail and expertise in special processes like chrome plating.

### The Issue Unfolds

During the audit, Rajesh scrutinized the **receiving inspection process** for outsourced components. SkyReach had entrusted PrimePlates Singapore, a specialist in chrome plating, with the surface treatment of undercarriage parts. These parts, critical for safety and durability, demanded flawless execution.

*However, Rajesh discovered a glaring gap. He issued a* ***major NC*** *under* ***clause 8.4.2 of AS9100D****, stating:*

*"Inspection to ensure that the externally provided process meets requirements is not effective. Verification activities of externally provided processes are to be performed according to the risks identified by the organization.*

*These shall include inspection or periodic testing, as applicable, when there is a high risk of NCs, including counterfeit parts. This is not evident from the below complaint: The customer has raised a complaint about the chrome plating of parts supplied to them."*

The NC rocked the team. The quality control staff, including Li Wei, the Inspection Lead, hesitated to accept responsibility. They had relied solely on the inspection of received parts, assuming that PrimePlates' certification guaranteed flawless work.

### The Challenge

The audit team demanded immediate containment and a detailed investigation into the NC. SkyReach's management, led by CEO Zhang Ming and Quality Director Mei Ling, recognized the gravity of the situation. Customer satisfaction was at risk, and their reputation in the aerospace industry hung in the balance.

### Intervention and Resolution

At first, Li Wei and his team resisted the suggestion of deeper involvement with the supplier's processes. However, Mei Ling stepped in with a composed yet assertive approach, stating: "This is not just about passing an audit. It is about our customers trusting us with their safety. We must ensure the integrity of every component, no matter where it is made."

Under her leadership, the team launched a three-phase action plan:

1. Containment:
   - **Customer Communication:** Immediate notification was sent to the customer acknowledging the complaint and outlining the steps being taken to resolve it.
   - **Inspection Enhancement:** SkyReach implemented 100% inspection of incoming chrome-plated parts until the root cause was addressed.
2. Root Cause Investigation:
   - A cross-functional team, including Process Engineer Wang Chen, visited PrimePlates Singapore to evaluate their chrome plating process.
   - The team identified a deviation in the temperature control of the plating bath, which led to uneven coating and occasional NCs.
3. Corrective Actions:
   - **Supplier Validation:** SkyReach established periodic validation of PrimePlates' special processes, including audits at their premises.
   - **Risk-Based Inspection:** Introduced risk-based inspection protocols for high-risk processes like chrome plating.
   - **Supplier Training:** Conducted training sessions for PrimePlates' staff to reinforce adherence to aerospace standards.

The actions were documented in a detailed **NCR**, including evidence of process changes, risk assessments, and the resolution of the customer complaint.

### Outcome and Acceptance

Rajesh reviewed the submitted NCR and found it met the stringent requirements of AS9100D. He acknowledged the efforts taken by SkyReach and approved the corrective actions. However, he noted that the effectiveness of these measures would be verified in the next surveillance audit.

The inclusion of supplier process validation as an ongoing activity was a gamechanger. SkyReach now had a stronger supplier management framework, ensuring critical processes met the required standards.

### Lessons Learned

This experience left a lasting impression on SkyReach Aviation:

1. Customer First: Every action, from inspection to corrective measures, must prioritize delivering a confirming product.
2. Proactive Supplier Management: Merely relying on supplier certifications is insufficient; ongoing validation and collaboration are vital.
3. Risk-Based Thinking: High-risk processes demand heightened scrutiny and periodic review.
4. Team Collaboration: Challenges are best tackled when management, engineering, and quality teams work together with a unified goal.

### Epilogue: A Stronger Future

A year later, during the next surveillance audit, the auditors observed a transformed organization. Processes were robust, and supplier performance had significantly improved. The corrective actions were found to be not only effective but also instrumental in preventing similar issues.

SkyReach Aviation emerged stronger from this experience, equipped with systems that guaranteed customer satisfaction and operational excellence. As Zhang Ming concluded during a company meeting: *"The sky is not the limit—it is our arena. Together, we'll ensure every part of every aircraft carries the stamp of SkyReach quality."*

## CS 64 [AS9100D Clause 8.4.3] Order Right or Regret: A Tale of Miscommunication

In the bustling city of Bengaluru, an organization manufacturing aircraft parts was gearing up for its initial certification audit. The air was thick with tension, but the auditor, a seasoned professional with a knack for storytelling, had a calming presence that transformed nervous faces into ones of curiosity and learning.

During the audit of the **purchase process**, the auditor noticed a recurring issue in the organization's approach to supplier communication. To lighten the mood and drive home an important point, he recounted a story that would soon become legendary within the company.

### The Tale of the Misplaced Order

"The production manager was in a bind," the auditor began, his voice calm but purposeful. "The next day's production depended on an urgent supply of raw materials. With little time to spare, he called the purchase manager, requesting immediate action.

The purchase manager, ever obliging, called the supplier and confidently placed the order. He repeated the requirements for clarity: '8×4 aluminum, 20 sheets, 20 gauge.' He even promised priority in payment. Relieved, he reassured the production manager, who, in turn, felt a wave of gratitude.

The next morning, the production manager eagerly awaited the truck at the material gate. But instead of a truck, a postman on a bicycle arrived, handing him an envelope. To his dismay, inside were **8 mm × 4 mm aluminum foils, 20-micron thickness, 20 pieces**—a far cry from the required material.

The production manager fumed momentarily but then realized the painful truth: **'If you order what you do not want, you'll get exactly what you do not want.'"**

The room burst into laughter, the tension dissipating. Yet, the lesson was clear, and the auditor's expression turned serious. "This story underscores the importance of clear, precise communication. Now, let's see whether your POs meet these standards."

## The Audit Findings

The auditor's thorough review revealed gaps in the organization's processes. POs lacked sufficient evidence of clear communication regarding product specifications, verification requirements, and logistics. He raised an NC against **clause 8.4.3—information for external providers**, emphasizing:

> *The organization shall ensure the adequacy of requirements before their communication to the external provider.*

The NC, though daunting at first, became a turning point. The purchase manager, along with the management representative and team, sprang into action.

1. **Containment Action:** Immediate checks were initiated on all ongoing POs to rectify errors and clarify requirements with suppliers.
2. **RCA:** A brainstorming session revealed a systemic issue—insufficient review protocols before issuing POs.
   - Process Drift: Over time, deviations from established procedures may have gone unnoticed.
   - Ineffective Monitoring: Lack of systematic audits or reviews to ensure compliance with defined conditions.
3. **Corrective Actions:**
   - A **checklist** was introduced to ensure completeness and clarity in all POs.
   - **Training sessions** were conducted to sensitize employees to the importance of precise communication.
   - The ERP system was updated to mandate verification steps before POs could be issued.

## A Cultural Shift

Two weeks later, the auditor reviewed the corrective action plan and evidence through OASIS, acknowledging the effort and accepting the plan. He assured the team he would verify the effectiveness of these measures during the surveillance audit the following year.

The auditor's tale became a regular point of discussion within the organization. Teams across departments began sharing their own anecdotes to highlight the importance of meticulousness. The story's moral, **"Order what you want, or get what you do not want,"** became an informal mantra.

## The Impact

The NC catalyzed a cultural and operational shift within the organization. Supplier management became more systematic, and procurement processes were streamlined. The production team reported fewer delays, and supplier relations improved significantly.

When the auditor returned a year later, he found a transformed organization. The purchase process, once riddled with inconsistencies, now stood as a model of efficiency and clarity.

As he concluded the surveillance audit, the auditor could not help but smile. "Sometimes," he mused, "all it takes is a story to rewrite an organization's story."

This emotional and instructive tale continues to inspire everyone in the organization to strive for excellence, ensuring that every order placed is precise, accurate, and effective.

## CS 65 [AS9100D Clause 8.4.3] Supplier Medals: Recognition of Excellence

### Scene 1: The Teacher's Eye

The conference room at Precision AeroSystems was bustling with activity as the AS9100D surveillance audit was underway. The audit team, led by Krishnan, a seasoned auditor with a penchant for teaching, was diving into the organization's supplier management processes.

Krishnan was known for his unique style. He did not just audit for compliance; he nurtured improvement. If something was good, he made it great. If it needed fixing, he did not hesitate to offer guidance.

As he reviewed the supplier performance evaluation records, he smiled. "This is impressive work," he said, addressing the Purchase Head, Meera. "Your evaluations are thorough, and you've identified your top-performing suppliers. But let me ask you something—what do you do for these high performers? Do you recognize their efforts?"

Meera hesitated. "We… do not have a formal recognition system. We share feedback occasionally, but nothing structured."

Krishnan leaned back in his chair, his tone thoughtful. "If the winner is not given a medal, they'll become just another face in the crowd. Excellence fades without recognition. Eventually, you'll be left with only the meek, weak, and disqualified to look at."

The room fell silent.

### Scene 2: The NC Is Issued

Krishnan opened the standard document, pointing to a clause. "Clause 8.4.3 clearly states that you must periodically review supplier performance, including conformity and OTD. But it also implies you must communicate this performance back to them. Right now, you're doing the evaluation but not closing the loop."

He handed Meera an NCR:

> ***NC 8.4.3:*** *Communication with suppliers does not include information about their performance.*
>
> ***Standard Reference:*** *Periodically review external provider performance and define necessary actions for improvement.*

"I'll help you overcome this small gap in your QMS," he said with a reassuring smile.

### Scene 3: Containment and RCA

The purchase team, led by Meera and supported by the management representative, immediately launched into action.

**1.** Containment:

- An initial communication was sent out to all active suppliers, thanking them for their contributions and informing them that a formal feedback system was being established.

**2.** RCA:

- **Root Cause:** The team identified a cultural gap—supplier evaluations were viewed as an internal activity rather than a two-way communication process.
- **Contributing Factors:**
  - Lack of awareness about the importance of supplier recognition.
  - No clear ownership of supplier communication in the evaluation process.

### Scene 4: Corrective Actions Implemented

The team worked collaboratively to implement corrective actions:

1. Supplier Recognition Program:
    - Introduced a formal system to recognize top-performing suppliers quarterly.
    - Created a "supplier excellence award" to be presented at an annual supplier meet.
2. Performance Feedback Reports:
    - Developed a standardized template to share performance feedback with all suppliers, highlighting areas of excellence and OFI.
3. Training and Awareness:
    - Conducted workshops for the procurement and supplier management teams on the importance of recognizing and engaging suppliers as partners.
4. Process Ownership:
    - Assigned a dedicated supplier relations manager to ensure regular communication and follow-up with suppliers regarding their performance.

### Scene 5: Closure of the NC

Three months later, Krishnan returned for a follow-up review. Meera presented the evidence of corrective actions:

- Supplier recognition event photographs and award lists.
- Feedback reports sent to all suppliers.
- Training session records.
- Revised supplier evaluation and communication procedures.

Krishnan reviewed the documents meticulously. "This is excellent progress," he said. "Your actions not only close the NC but also strengthen your relationship with suppliers. This will pay off in the long run."

He signed off on the NC closure and added the new supplier engagement process to the agenda for the next surveillance audit.

### Epilogue: The Power of Recognition

Months later, Precision AeroSystems noticed a marked improvement in supplier performance. Suppliers who received feedback and recognition became more committed, while those who needed improvement responded positively to the guidance.

As Meera reflected on the experience, she realized Krishnan's wisdom. "A little recognition goes a long way," she said to her team. "Our suppliers aren't just vendors—they're partners in our success. It is our job to keep them motivated."

The lesson was clear: in quality management, building partnerships is as important as maintaining processes. Recognizing excellence fosters loyalty, drives improvement, and creates a culture of mutual growth.

## CS 66 [AS9100D Clause 8.4.3] Supplier Silence: Communication's Role in Aerospace Success

Hyderabad-based Avionics Software Solutions, a budding enterprise in the aerospace domain, was on the brink of achieving a prestigious milestone: certification for AS9100. This certification was their gateway to collaborating with global aerospace giants, a dream the company had nurtured since its inception.

Leading up to the certification audit, the company's Purchase Manager, Sampath Kumaran, had meticulously designed a system to manage externally provided processes, products, and services. His system aimed to align with the rigorous quality expectations of the aerospace sector. However, during the audit, the unexpected happened.

The auditor flagged an NC under **clause 8.4.3** regarding supplier communication and performance monitoring:

1. Supplier communication lacked a mechanism to raise awareness among suppliers' personnel about:
    - Their contribution to product or service conformity.
    - Their role in ensuring product safety.
    - The importance of ethical behavior.
2. The organization did not monitor supplier performance against benchmarked standards. POs were being issued without considering suppliers' past performance.

This revelation came as a surprise to Kumaran and the team, who had believed their processes were robust.

### Follow-Up Actions

The company sprang into action to address the gaps identified:

1. Enhanced Supplier Communication:
    - A detailed supplier awareness program was introduced. Suppliers were provided with documentation and periodic training on:
        - Their role in maintaining product and service quality.
        - The criticality of safety in aerospace applications.
        - Ethical practices aligned with industry norms.
    - These requirements were explicitly stated in all new supplier agreements and POs.
2. Supplier Performance Monitoring:
    - A supplier performance evaluation system was implemented. This system included metrics for delivery timelines, defect rates, and compliance with quality requirements.
    - A centralized dashboard was introduced for real-time monitoring of supplier performance.
    - Only suppliers meeting or exceeding benchmarked standards were approved for new POs.
3. Process Documentation and Validation:
    - Comprehensive documentation of these corrective actions was prepared and reviewed.

- Internal training sessions were held to ensure that all team members understood and adhered to the new processes.

The auditor reviewed the corrective actions and accepted them as satisfactory. However, the verification of the effectiveness of these actions was scheduled for the next surveillance audit.

### A Lesson Learned

This experience was a wake-up call for Avionics Software Solutions. It underscored the critical role of communication, not just within the company but with external partners as well. The team recognized that clear, consistent communication and robust monitoring were essential to meet the exacting standards of the aerospace industry.

This incident ultimately strengthened the organization's commitment to continual improvement, setting the stage for a successful journey in the aerospace sector.

## CS 67 [AS9100D Clause 8.4.3] The Forgotten Clause: A Lesson in Intellectual Property Security

### Scene 1: A Misstep Unnoticed

At RotorWings Aerospace, a renowned manufacturer of helicopters, the procurement team prided itself on its seamless relationships with suppliers. Over the years, they had built trust with their vendors, sharing technical drawings, process specifications, and proprietary designs to ensure quality in outsourced manufacturing.

One sunny afternoon, Anjali, the Quality Manager, sat in her office reviewing the company's procurement processes in preparation for the upcoming AS9100D surveillance audit. Everything appeared in order until she skimmed the terms and conditions of their POs.

"Suppliers must return intellectual property after use," she read aloud. "Interesting clause... but have we ever ensured compliance?"

Anjali made a mental note to revisit the issue but got sidetracked with other pressing tasks.

### Scene 2: The Audit Begins

During the audit, Gowda, a meticulous auditor with a reputation for uncovering hidden risks, reviewed RotorWings' procurement processes.

While examining the supplier management records, he asked casually, "How do you ensure that suppliers return intellectual property—drawings, specifications, and other sensitive documents—after completing their tasks?"

The Procurement Head, Suresh, blinked. "Well, it is in the PO terms and conditions. We trust our suppliers to follow it."

Gowda's brow furrowed. "Trust is important, but verification is critical. Have you ever checked whether these documents are returned?"

Suresh hesitated. "No, we haven't… but none of our suppliers have ever misused them."

Gowda leaned forward. "How can you be sure? Intellectual property theft doesn't always make noise when it happens. Let me share a story."

### Scene 3: A Cautionary Tale

Gowda recounted a real-life incident:

"A few years ago, a defense manufacturer shared proprietary designs with a supplier to produce critical components. Those designs found their way into the hands of a competitor, who reverse-engineered the technology. The competitor launched a similar product at a fraction of the cost, devastating the original company's market share. Investigations revealed that the supplier hadn't returned the drawings, which were later leaked to an unauthorized third party. The financial and reputational damage was irreversible. Two employees were taken into custody last year in another company by police, after recovering some intellectual property from sensitive areas."

He paused for effect, then added, "RotorWings is dealing with helicopters—technology that has both commercial and defense implications. Can you afford such a risk?"

Anjali, sitting nearby, felt a chill. "We need to act on this immediately."

### Scene 4: The NC

Gowda issued the NC:

***Clause 8.4.3 Violation:*** *No process to ensure suppliers return intellectual property after use, as stipulated in the PO terms and conditions.*

***Evidence:*** *A review of supplier records showed no documented confirmations or tracking of returned intellectual property.*

### Scene 5: Follow-Up Actions

Anjali and her team launched an immediate containment plan and corrective actions:

1. Containment:
    - Issued an urgent communication to all active suppliers, requesting the return of any drawings, specifications, or intellectual property in their possession.
2. RCA:
    - Root Cause: Over-reliance on trust and no established process for tracking the return of intellectual property.
    - Contributing Factors: Lack of awareness among procurement staff about the criticality of intellectual property security.
3. Corrective Actions:
    - Process Update:
        - Incorporated a mandatory clause in POs requiring suppliers to sign an acknowledgement of intellectual property receipt and return.
        - Established a tracking system to monitor intellectual property provided to and returned by suppliers.
    - Training: Conducted training for the procurement and quality teams on

intellectual property risks and the importance of enforcing return policies.
- Audit Mechanism: Introduced periodic audits of supplier compliance with intellectual property return requirements.

### Scene 6: Consequences Realized

Weeks into the process, one supplier admitted they had retained multiple drawings for convenience. "We didn't realize it was such a big deal," they said.

Anjali shook her head. "It is not just a big deal—it is a potential disaster. We'll ensure this doesn't happen again."

### Scene 7: Verification and Closure

A year later, during the next surveillance audit, Gowda returned.

He reviewed the updated process and checked supplier records. Each supplier now had signed acknowledgments for intellectual property returns, and the tracking system showed a 100% compliance rate.

Gowda nodded in approval. "RotorWings has taken a significant step forward in safeguarding its intellectual property. This is not just compliance—it is smart business."

Anjali smiled, relieved but resolved. "We learned our lesson the hard way. Intellectual property is the backbone of our innovation. Protecting it is non-negotiable."

### Epilogue: A Culture of Vigilance

RotorWings Aerospace emerged stronger from the experience. The corrective actions fostered a culture of vigilance and accountability, ensuring that their intellectual property remained secure.

What is the lesson? Trust is good, but a robust process is better. In the high-stakes world of aerospace manufacturing, intellectual property is as valuable as the technology it creates and must be protected at all costs.

## CS 68 [AS9100D Clause 8.4.3] Breaking the Silos: A Lesson in Quality and Communication

### Scene 1: The Silos Within

At SkyReach Aeronautics, one of the world's largest aircraft manufacturers, the production floor buzzed with activity. From design to assembly, departments operated like independent kingdoms, each focused on its own objectives.

Sarah, the company's Management Representative, often noticed the lack of communication between teams. Quality concerns addressed in one department rarely reached others. "We need to fix this," she muttered to herself during a routine meeting, but the pace of daily operations left little room for systemic changes.

### Scene 2: The Audit Begins

When the AS9100D certification audit commenced, the team of auditors, led by veteran auditor **Patel**, dove deep into the organization's processes.

Patel, known for his thoroughness, made his rounds, gathering insights from his colleagues on their respective findings. One auditor, Priya, reviewed the **customer defect reports** and noticed a recurring issue: several parts had failed during customer operations. Investigations pointed to the **poor quality of raw materials**, particularly for critical components.

Meanwhile, another auditor, Ramesh, observed a glaring gap in the **receiving inspection area**. "The team accepts raw materials based solely on supplier-provided test reports," he noted in his findings. "There's no process to revalidate the test data, even for critical items."

During the mid-audit debrief, Patel pieced these findings together. "We might be looking at a systemic issue," he said, his brows furrowed.

### Scene 3: The NC Revelation

In the final meeting with Sarah, Patel laid out the audit findings.

Patel: "Sarah, the standard's clause 8.4.3 is explicit: when relying on supplier test reports for critical items, organizations must evaluate the data and validate its accuracy. SkyReach failed to do this. Here's an example: the material used in a critical part failed due to poor quality, as identified in the defect investigation report. Yet, no action was taken to improve the material verification process."

Sarah: "But we trust our suppliers! Their test reports are certified."

Patel: "Trust doesn't replace verification, especially for critical items. The standard doesn't just ask you to believe the data—it asks you to confirm it. This failure is not isolated. It is systemic, stemming from siloed operations and a lack of cross-departmental communication."

He issued the NC:

***Clause 8.4.3 Violation:*** *No process to validate the accuracy of supplier test reports for critical materials.*

***Evidence:*** *Customer defect report findings, defect investigation reports, and observations from the receiving inspection area.*

### Scene 4: Follow-Up Actions

Determined to address the NC, Sarah led a cross-departmental team to undertake the following corrective actions:

**1.** Containment:
- Halted the production of parts involving raw materials from the identified supplier.
- Conducted an immediate revalidation of test reports for all raw materials in stock.

**2.** RCA:
- The team identified two primary causes:
  - Over-reliance on supplier-provided test reports without verification.
  - Lack of communication between defect investigation and receiving inspection teams.

**3.** Corrective Actions:
- Implemented a process to revalidate supplier test reports for all critical items.
- Created a communication protocol to ensure defect investigation findings are shared across relevant departments.
- Trained employees in the receiving inspection and quality assurance teams on the updated process.

**4.** Preventive Measures:
- Established a centralized database for defect investigation findings and quality concerns, accessible to all departments.
- Introduced regular cross-departmental review meetings to share lessons learned and best practices.

### Scene 5: Audit Closure

Three weeks later, Sarah submitted the **NCR** with evidence to Patel.

Patel reviewed the documents carefully:

- Updated receiving inspection procedures.
- Records of revalidated test reports.
- Training session logs.
- Implementation of the centralized defect database.

He nodded in approval. "The corrective actions are sound. I'll close the NC for now, but I'll verify the effectiveness during the next surveillance audit."

Sarah exhaled in relief but resolved to ensure the new processes became a part of SkyReach's culture.

### Epilogue: Breaking the Silos

Months later, during a routine quality meeting, Sarah noticed a change. Teams were sharing insights freely, and the centralized database had become an indispensable tool. The once-isolated silos were dissolving, replaced by a collaborative environment.

The lesson? In a complex organization, quality is not just about processes—it is about communication, accountability, and a shared commitment to excellence. SkyReach Aeronautics had learned the hard way, but they were stronger for it.

## CS 69 [AS9100D Clause 8.5.1 a to q] Audit Challenges in Manufacturing: Turning Gaps into Triumphs

In the bustling industrial hub of Pune, India, a medium-scale manufacturing unit specializing in aeronautical parts was undergoing a critical audit. The stakes were high: the company aimed to establish itself as a reliable supplier to defense customers. The Head of Production, Rajesh Malhotra, was at the forefront, tasked with addressing the findings and ensuring a robust production process that met stringent aerospace standards.

### Audit Findings: Identifying the Gaps

The audit progressed smoothly until the auditors uncovered a significant NC under AS9100D clause 8.5.1: controlled conditions were not adequately established and monitored in the production and service provision processes. The listed gaps spanned a range of crucial elements, including:

- Absence of detailed product information and quality characteristics.
- Missing or incomplete drawing specifications and process documentation.
- Lack of defined criteria for acceptance and rejection.
- Unestablished verification operations in sequence.
- No formal statistically valid sampling plan.
- Insufficient controls over critical items and in-process inspections.
- Inadequate periodic validation of special processes.
- Weak controls for utilities and supplies.

- Lack of mechanisms to prevent, detect, and remove foreign objects.
- Incomplete tracking and traceability of products for recall or replacement.

These deficiencies posed a risk to product quality and compliance, jeopardizing the company's aspirations to enter the aerospace defense market.

### Rajesh's Determination

Rajesh Malhotra understood the gravity of the situation. He convened an emergency meeting with his team, including key members from quality assurance, engineering, and production planning. "This is not just about addressing an audit finding," he emphasized. "It is about building a production process that earns the trust of our customers and ensures our products meet the highest standards of safety and precision."

### The Action Plan

Rajesh outlined a phased approach:

1. Comprehensive Documentation:
    - Collaborated with the marketing team to ensure detailed capture of customer requirements, including drawings, specifications, and expected results.
    - Standardized templates for recording quality characteristics, acceptance/rejection criteria, and process validation checkpoints.
2. Enhanced Training and Competency:
    - Identified critical competencies required for special processes.
    - Organized training sessions to upskill personnel and ensure certification for specialized tasks.
3. Process and Tool Standardization:
    - Established clear instructions for the use of measurement equipment and tools.
    - Implemented product-specific software programs for design validation and process simulation.
4. Validation and Monitoring:
    - Developed a statistically valid sampling plan to ensure consistent quality.
    - Scheduled periodic validations for special processes and utilities.
5. Human Error Prevention:
    - Implemented foolproof systems, including checklists and automated alerts, to minimize human error. Sometimes methodology known as the Swiss cheese model is applied. The Swiss cheese model of accident causation is a model used in risk analysis and risk management. It likens human systems to multiple slices of Swiss cheese, which have randomly placed and sized holes in each slice, stacked side by side, in which the risk of a threat becoming a reality is mitigated by the differing layers and types of defenses which are "layered" behind each other. Therefore, in theory, lapses and weaknesses in one defense do not allow a risk to materialize since other defenses also exist (e.g., other slices of cheese), to prevent a single point of failure.
6. End-to-End Traceability:
    - Introduced a robust tracking system to monitor product accountability at every stage of production.
    - Maintained detailed records of in-process inspections and final releases.
7. Foreign Object Prevention:
    - Instituted regular checks and cleaning procedures to prevent, detect, and remove foreign objects.

### Corrective Action Implementation

Over the next month, Rajesh's team worked tirelessly to close the gaps. They conducted internal audits to verify the implementation of controlled conditions, ensuring every aspect from clause 8.5.1 (a) to (q) was addressed with tangible evidence.

### Highlights of Corrective Actions

- **Workmanship Criteria:** Clear guidelines were established, with visual aids for acceptable and unacceptable standards.
- **In-Process Inspections:** Verification points were defined, and signoffs were mandated at critical stages.
- **Utility Controls:** Monitoring systems were installed to ensure consistent supply quality.
- **Post-Delivery Activities:** Procedures were developed for tracking and managing post-delivery customer feedback and complaints.
- **Periodic Validation:** Special processes were validated at scheduled intervals to maintain reliability.

The management representative meticulously documented the corrective actions and submitted the NCR format to the auditor.

### Auditor's Review

When the auditor reviewed the corrective actions and supporting evidence online, they were impressed by the thoroughness and speed of implementation. "This is a textbook example of turning an audit finding into an opportunity for excellence," the auditor remarked. The team's efforts were validated with the recommendation for initial certification.

The auditor also emphasized the importance of sustaining these improvements. A follow-up audit was scheduled for the next year to verify the effectiveness of the actions.

### Preparing for the Future

Rajesh knew the journey did not end with the certification. He established a culture of periodic internal audits to ensure ongoing compliance. "This is not just about passing an audit," he reiterated to his team. "It is about building a legacy of quality and trust in the aerospace industry."

With the foundation of controlled conditions firmly in place, the company was ready to take on the challenge of supplying aerospace parts to defense customers, confident in their ability to deliver excellence.

Chapter 07

# Audit Nonconformities in Product Realization Processes—What Can Happen If Controlled Conditions Are Not Maintained

Let us give a break to the stories and come to a specific point. Let us analyze what happens if the specified controlled conditions numbered 8.5.1 "a" to "q" are not maintained during production.

## CS 70 [AS9100D Clause 8.5.1 a] Certification Audit Story: The Turning Point at Stellar Aerospace

Once upon a lively season at Stellar Aerospace Ltd., the team was buzzing with excitement—it was finally time for their certification audit for AS9100D. Months of preparation, late evenings, and team workshops had built up to this grand event. The entire plant wore a polished look, and everyone was ready to showcase their commitment to quality excellence.

Enter Mr. Arjun Singh, the lead auditor—a sharp, experienced professional known for his detailed eye and encouraging attitude. After a warm welcome and opening meeting, the audit trails began: document reviews, interviews, and plant walkarounds.

As Mr. Singh strolled through the production floor, everything seemed in order—skilled operators, advanced equipment, and neatly stacked products. But when he asked to see the work instructions or documented specifications for a new product line, the team hesitated. There were no formal documented specifications/drawings or instructions readily available—only verbal directions and memory-based actions were taken!

### Audit Finding

*Controlled conditions were not maintained in the production area due to the absence of documented information defining:*

- *The characteristics of the products to be produced, the services to be provided, or the activities to be performed.*
- *The results to be achieved.*

Without proper documented information, the risk of errors, inconsistencies, and customer dissatisfaction was significantly elevated.

Arjun, while empathetic, had no choice but to raise an NCR—a wake-up call for Stellar Aerospace.

### Correction/Immediate Containment Actions

- Document Review Initiated: A task force quickly gathered all available documentation and identified missing or outdated items.
- Controlled Production Halt: Production on affected lines was paused or moved to a controlled mode until proper documentation was developed.
- Temporary Work Instructions: Clear, concise temporary instructions were issued immediately to ensure activities remained consistent.
- Inspection and Testing: All in-progress and finished items were inspected thoroughly to validate conformity to requirements.

Containment was achieved within three working days.

### RCA

Using a 5 Whys analysis, the team uncovered the underlying issues:

- Primary Cause: Lack of responsibility for document updates during product changes.
- Contributing Factors:
  - No formalized linkage between change management and document control.
  - Training gaps regarding the importance of documented work instructions.
  - No regular internal audits focused on document control effectiveness.

### Short-Term Corrective Actions

- Complete Documentation Update: All product specs, process instructions, and acceptance criteria were documented and uploaded into a centralized database (implemented within 30 days).
- Training and Awareness Sessions: All operators and supervisors underwent mandatory training on document access and use (completed alongside documentation update).
- Appointment of a Document Control Officer: A dedicated officer was assigned to own, manage, and maintain document control (position filled within 20 days).

### Long-Term Corrective Actions

- Document Control Process Enhancement: New procedures were put in place for document creation, revision, approval, and distribution, with proper version controls.
- Scheduled Internal Audits: Quarterly internal audits focusing on document availability and usage were introduced.

- Management Review Integration: Documentation compliance became a regular review item at management review meetings.

Long-term actions were fully implemented within 90 days.

### Effectiveness Verification of Corrective Actions

- Follow-Up Internal Audits: Surprise internal audits showed full compliance with clause 8.5.1 (a)—no missing documents, clear work instructions at the required workstations.
- Production and Quality Metrics:
  - Product defect rates decreased by 30%.
  - Rework instances dropped by 25% over the next two quarters.
- Employee Surveys: Post-training surveys showed a 40% increase in operator confidence and understanding of process requirements.
- Customer Feedback: There were no customer complaints related to production inconsistencies for two consecutive quarters.

Effectiveness was fully verified within 120 days after NCR closure.

### Benefits Derived by Stellar Aerospace

The NC, though initially disappointing, became a blessing in disguise. Through disciplined corrective actions, Stellar Aerospace achieved:

1. Improved Process Control: Clear and accessible documented work instructions reduced process variability.
2. Enhanced Product Quality: Consistency and compliance across all product lines improved drastically.
3. Stronger Employee Engagement: Workers felt empowered with proper tools and clear guidance.
4. Operational Efficiency: Reduced rework, faster training for new staff, and smoother internal communications.
5. Customer Trust Reinforced: Improved delivery performance and quality assurance led to stronger, longer-term client relationships.
6. Future-Readiness: With a robust document control system and a proactive culture, Stellar Aerospace was now confidently poised to scale new heights in the aerospace industry.

And when Mr. Singh returned for the final certification /corrective action review, he smiled broadly, declaring: “Congratulations, Stellar Aerospace—you have successfully demonstrated compliance with AS9100D. Certification well deserved!”

From a minor stumble to a major leap forward, the journey had made Stellar Aerospace stronger than ever!

## CS 71 [AS9100D Clause 8.5.1 b] Uncontrolled Conditions: Lacking Proper Monitoring Resources

*Controlled conditions for production process—b. Controlled conditions are not maintained for the availability and use of suitable monitoring and measuring resources.*

### Containment Action

Immediately mitigate the impact of the NC on production or service provision.

- **Identify Affected Resources:** Review all monitoring and measuring equipment used in the production process. Identify any equipment that is unavailable, unsuitable, or out of calibration.
- **Segregate Nonconforming Outputs:** Quarantine or re-inspect any products or services that were tested or measured using unsuitable equipment to confirm compliance with specifications.
- **Provide Temporary Resources:** Arrange alternative, verified monitoring and measuring equipment to continue operations under controlled conditions.
- **Communicate to Teams:** Inform relevant personnel about the issue and ensure interim controls are implemented immediately.

### RCA

Identify the underlying reason(s) for the lack of availability or use of suitable monitoring and measuring resources. Possible causes:

- **Lack of Equipment Maintenance:** Monitoring and measuring equipment was not maintained, leading to failures or unavailability.
- **Calibration Oversight:** Equipment used without proper or up-to-date calibration records.
- **Inventory Issues:** Insufficient monitoring/measuring tools or delays in procurement of required resources.
- **Process Gaps:** Ineffective planning or scheduling for equipment calibration and verification.
- **Training Deficiencies:** Personnel were unaware of the requirements for using suitable monitoring and measuring resources.

### Techniques for Analysis

- **5 Whys:** Repeatedly ask "Why?" to identify the root cause.
- **Fishbone Diagram:** Analyze potential causes under categories such as equipment, process, people, materials, and management.
- **Audit Review:** Examine records and audit reports for evidence of systemic issues in equipment management.

### Short-Term Corrective Actions

1. **Inspect and Calibrate:** Verify and calibrate all monitoring and measuring equipment immediately. Replace or repair any faulty or unsuitable equipment.
2. **Document and Communicate Standards:** Ensure all personnel are aware of the standards and requirements for monitoring and measuring resources.
3. **Temporary Workarounds:** If equipment cannot be replaced immediately, outsource measurements to a qualified external service provider.

### Long-Term Corrective Actions

1. Implement an Equipment Management System:
   - Maintain an inventory of all monitoring and measuring equipment with detailed records.
   - Use a tracking system for calibration schedules, status, and service history.
2. Develop a Calibration Policy: Establish clear procedures for calibration, including periodic verification, record-keeping, and ensuring traceability to national/ international standards.
3. Train Personnel: Provide regular training to operators and maintenance staff on the importance of using suitable monitoring and measuring resources.
4. Audit and Review:
   - Incorporate checks for monitoring and measuring resources into internal audit programs.
   - Use KPIs to monitor equipment compliance rates.
5. Procurement Planning: Create a process to assess future needs for monitoring and measuring equipment and ensure timely procurement.

### Example of Corrective Action Statement

"We will establish a centralized equipment management system to track the availability, suitability, and calibration status of monitoring and measuring resources. This includes implementing a preventive maintenance schedule, assigning responsibility for equipment verification, and conducting regular audits to ensure compliance. Additionally, we will train employees on proper use and requirements for monitoring resources."

### Preventive Measures

- Automation: Use digital tools or software to send reminders for calibration and maintenance due dates.
- Risk Assessment: Periodically assess risks associated with equipment failure or unavailability.
- Continual Improvement: Gather feedback from audits and NCs to refine equipment management processes.

## CS 72 [AS9100D Clause 8.5.1 c] Uncontrolled Conditions: Monitoring and Measurement Gaps

*Controlled conditions for production process clause 8.5.1—C. Controlled conditions are not maintained for the implementation of monitoring and measurement activities at appropriate stages to verify that criteria for control of processes or outputs, and acceptance criteria for products and services, have been met, including criteria for acceptance.*

### Containment Action

- Identify Affected Processes and Products: Review the production processes to identify stages where monitoring and measurement are missing or not adequately implemented. Identify any outputs (products or services) affected by this gap.
- Stop Production or Segregate Outputs: Halt production or service provision for processes lacking defined monitoring and measurement criteria. Segregate and inspect completed products or in-process outputs to ensure they meet acceptance criteria.
- Implement Temporary Monitoring: Introduce immediate checks (manual or automated) at critical stages to verify process control and product/service compliance.
- Notify Relevant Personnel: Communicate the issue to all relevant departments and teams to ensure temporary measures are strictly followed.

### RCA

Possible causes:

- Process Design Flaws: Monitoring and measurement stages not adequately defined during process design or planning.
- Lack of Criteria Definition: Acceptance criteria for processes, outputs, or products not clearly documented or communicated.
- Training Deficiency: Personnel unaware of when and how monitoring and measurement should be implemented.
- Resource Constraints: Insufficient equipment, tools, or personnel for conducting monitoring and measurement activities.
- Ineffective Process Review: Inadequate review or auditing of process controls to ensure monitoring and measurement criteria are met.

### Tools for RCA

- 5 Whys: Ask "Why?" repeatedly.
- Process Mapping: Visualize the process flow to identify missing monitoring and measurement stages.
- Pareto Analysis: Focus on the most significant gaps contributing to this issue.

### Corrective Action

Objective: To address the root cause and ensure monitoring and measurement activities are implemented effectively and consistently.

### Short-Term Actions

- Define Immediate Monitoring Points:
  - Identify critical stages in the process where monitoring and measurement are essential and ensure they are immediately implemented.
  - Establish temporary acceptance criteria where needed.

- Conduct Immediate Inspections: Review and retest all outputs produced during the affected period to confirm compliance with product/service requirements.
- Communicate Criteria: Share acceptance criteria and monitoring protocols with all relevant personnel.

### Long-Term Actions

- Revise Process Documentation: Update process control documents (e.g., control plans, work instructions, inspection checklists) to include clear monitoring and measurement activities at appropriate stages.
- Define acceptance criteria for processes, products, and services in detail.
- Develop a Monitoring Plan: Create a monitoring and measurement plan that specifies the following:
    - Stages where monitoring and measurement are required.
    - Methods, tools, and frequency of monitoring activities.
    - Acceptance criteria for each stage.
- Enhance Training Programs: Train employees on the updated monitoring and measurement procedures, focusing on their roles and responsibilities.
- Implement Real-Time Monitoring Tools: Where feasible, integrate automated systems or software for real-time monitoring and measurement of process outputs.
- Strengthen Audits and Reviews: Incorporate regular internal audits to ensure monitoring and measurement activities are being conducted as per defined standards.
- Use findings from audits to improve processes continually.

### Preventive Measures

- Process Validation: Conduct periodic validation of processes to ensure monitoring and measurement stages remain effective.
- Continual Feedback Loop: Collect and analyze feedback from production teams and quality audits to address emerging gaps proactively.
- Management Oversight: Include the monitoring plan in management review meetings to ensure sustained attention and support.

### Example of Corrective Action Statement

"We will revise the process documentation to define and implement monitoring and measurement activities at all critical stages, with clear acceptance criteria for processes and outputs. Training sessions will be conducted for all relevant personnel to ensure consistent implementation. Regular audits and a digital monitoring system will be established to verify compliance and sustain process control."

## CS 73 [AS9100D Clause 8.5.1 d] Inadequate Control: Gaps in Infrastructure and Process Environment

*Controlled conditions for production process 8.5.1 d. Controlled conditions are not maintained for the use of suitable infrastructure and environment for the operation of processes.*

### Containment Action

Immediate steps to mitigate the issue and prevent further occurrences in the short term

1. Isolate Affected Activities or Products:
   - Halt all operations involving undocumented or improperly documented processes/products.
   - Notify relevant stakeholders about the issue.
2. Provide Interim Documentation:
   - Generate temporary documented information that outlines critical characteristics and requirements for the affected products, services, or activities.
3. Implement a Temporary Control System:
   - Assign a team to oversee adherence to interim guidelines until full resolution.
   - Enhance supervision and validation steps during production or service provision.

### RCA

Identify the underlying cause of the issue:

1. Assess Procedural Gaps:
   - Review existing policies for documenting product characteristics, services, and activities.
   - Determine whether the issue arose from incomplete or unclear documentation requirements.
2. Examine Training and Awareness:
   - Evaluate whether employees were adequately trained to understand and maintain controlled conditions.
   - Identify knowledge gaps related to the importance of documentation.
3. Evaluate Document Control Systems:
   - Inspect whether the current system for creating, maintaining, and updating documented information is effective.
   - Check for potential software or process failures leading to missing documentation.
4. Investigate Communication Channels:
   - Determine whether there were miscommunications or lack of accountability for documentation tasks.

### Corrective Action

Steps to permanently address the root cause and prevent recurrence:

1. Develop or Revise Documented Procedures:
   - Create or update standard operating procedures (SOPs) to define required documented information clearly.
   - Ensure procedures specify roles, responsibilities, and timelines for documentation.

2. Enhance Training Programs:
   - Conduct training sessions for employees on the importance of controlled conditions and proper documentation.
   - Include real-world scenarios and examples to improve understanding.
3. Implement Robust Document Control Systems:
   - Adopt an automated document management system to ensure consistency, version control, and accessibility.
   - Introduce alerts for missing or outdated documentation.
4. Regular Audits and Monitoring:
   - Establish routine audits to verify compliance with documentation requirements.
   - Use audit findings to identify and address systemic issues.
5. Accountability Measures:
   - Assign specific personnel to oversee the documentation process.
   - Create a feedback loop to address issues immediately and improve continually.
6. Communication Improvements:
   - Enhance communication channels to ensure clear expectations for documentation and reporting issues.

## CS 74 [AS9100D Clause 8.5.1 e] Uncontrolled Conditions: Gaps in Competent Personnel Appointments

*Controlled conditions for production process 8.5.1 e. Controlled conditions are not maintained for the appointment of competent persons, including any required qualification.*

### Containment Action

1. Review Current Personnel Assignments:
   - Identify positions filled by individuals who lack the required qualifications or competence for their roles.
   - Conduct an immediate skills and qualifications audit for key positions.
2. Reassign Responsibilities:
   - Temporarily reassign tasks to qualified personnel or supervisors while addressing gaps.
   - In critical areas, halt operations where unqualified personnel could pose safety or quality risks.
3. Provide Immediate Training or Support:
   - Offer short-term training, mentoring, or guidance to address immediate skill gaps.
   - Pair less competent employees with experienced staff for oversight.
4. Notify Stakeholders:
   - Inform management and relevant teams about the issue and interim measures being implemented.

## RCA

1. **Evaluate Recruitment Processes:** Review hiring practices to determine whether qualification and competency requirements were clearly defined and followed. Check whether job descriptions and competency criteria were inadequate or outdated.
2. **Assess Training and Onboarding Programs:** Investigate whether new hires or promoted employees received sufficient training or onboarding for their roles.
3. **Analyze Competency Verification Processes:** Examine how the organization verifies qualifications, certifications, and skills during recruitment or internal promotions.
4. **Identify Resource Constraints:** Determine whether the issue resulted from insufficient staffing, budget constraints, or pressure to fill roles quickly.
5. **Investigate Communication Gaps:** Check whether there was a breakdown in communication regarding qualifications and competency requirements between departments or hiring managers.

## Corrective Action

1. Define and Document Competency Requirements:
   - Develop or update job descriptions with clear qualifications, certifications, and skill requirements for each role.
   - Include measurable criteria for assessing competence.
2. Strengthen Recruitment and Selection Processes:
   - Implement robust screening mechanisms to verify qualifications, certifications, and prior experience.
   - Use structured interviews and competency-based assessments during hiring.
3. Enhance Training and Development Programs:
   - Introduce comprehensive onboarding programs to equip new hires with role-specific skills.
   - Provide ongoing training and certification opportunities for employees to maintain or enhance competence.
4. Establish a Competency Verification System:
   - Create a system to periodically review and validate employee qualifications and skills.
   - Maintain records of certifications, licenses, and training for all employees.
5. Implement a Performance Monitoring Framework:
   - Use KPIs to regularly assess the effectiveness of personnel in their roles.
   - Provide feedback and additional support as needed.
6. Improve Communication and Collaboration:
   - Facilitate better coordination between HR, department heads, and management to align on competency requirements.
7. Resource and Succession Planning:
   - Ensure adequate staffing levels and plan for succession in key roles to avoid rushed or unqualified appointments.
   - Develop internal talent pools to prepare employees for future roles.
8. Conduct Periodic Audits:
   - Regularly audit recruitment, training, and competency management processes to identify and correct gaps.

## CS 75 [AS9100D Clause 8.5.1 f] Validation Gaps: Uncontrolled Conditions in Special Processes

*Controlled conditions for production process—controlled conditions are not maintained for the validation, and periodic revalidation, of the special processes.*

### Containment Action

1. Identify Affected Processes:
   - Review all ongoing special processes to determine which ones lack initial validation or periodic revalidation.
   - Segregate or quarantine any outputs from unvalidated or expired processes to prevent nonconforming products or services from being delivered.
2. Temporarily Halt Unvalidated Processes:
   - Suspend operations for the processes that require validation or revalidation until proper evaluation is conducted.
3. Conduct Emergency Validation:
   - Perform expedited validation of critical special processes to ensure compliance and product integrity.
   - Use documented procedures and qualified personnel for this temporary measure.
4. Enhance Supervision and Oversight:
   - Assign qualified personnel to monitor special processes closely and verify outcomes.
5. Communicate with Stakeholders:
   - Notify management, quality teams, and customers (if necessary) about the issue and measures being taken to ensure compliance.

### RCA

1. Review Validation Policies:
   - Check whether there were clear and documented procedures for validating and revalidating special processes.
   - Evaluate whether these procedures were followed consistently.
2. Examine Training and Competence:
   - Investigate whether personnel responsible for validation understood the requirements and importance of periodic revalidation.
   - Determine whether there were any knowledge or skill gaps.
3. Analyze Monitoring Systems:
   - Assess whether there were mechanisms in place to track and schedule revalidation activities.
   - Check whether oversight or monitoring systems failed to flag overdue revalidation.
4. Evaluate Resource Allocation:
   - Investigate whether resource constraints (personnel, time, budget) contributed to the failure to validate or revalidate processes.
5. Investigate Past Performance:
   - Review previous validation and revalidation cycles for signs of systemic issues, such as delays, incomplete documentation, or missed deadlines.

### Corrective Action

1. Develop or Update Validation Procedures:
   - Create or revise detailed procedures for the initial validation and periodic revalidation of all special processes.
   - Clearly define roles, responsibilities, and criteria for successful validation.

**2.** Implement a Tracking System:
- Introduce a system (e.g., software, logs) to monitor and schedule validation and revalidation activities.
- Set automated alerts to notify responsible personnel of upcoming or overdue validations.

**3.** Enhance Training Programs:
- Train employees and supervisors on the importance of validating special processes and adhering to revalidation schedules.
- Include guidelines on recognizing and reporting deviations.

**4.** Perform Comprehensive Revalidation:
- Conduct a full-scale revalidation of all special processes to ensure compliance with current standards.
- Document results and adjust processes as necessary.

**5.** Strengthen Documentation Practices:
- Ensure all validations and revalidations are thoroughly documented, including test methods, acceptance criteria, and results.
- Maintain records for auditing and traceability purposes.

**6.** Introduce Audits and Reviews:
- Schedule periodic internal and external audits to verify compliance with validation requirements.
- Use audit findings to improve processes and address systemic gaps.

**7.** Allocate Resources:
- Ensure adequate personnel, time, and budget are available to support validation activities.
- Assign dedicated personnel to oversee validation and revalidation programs.

**8.** Collaborate with Stakeholders:
- Work with process owners, quality teams, and external experts (if necessary) to ensure all special processes are validated and meet industry standards.

**9.** Monitor Performance Continually:
- Use KPIs to track the effectiveness of validation processes and identify areas for improvement.

## CS 76 [AS9100D Clause 8.5.1 g] Uncontrolled Conditions: Lack of Human Error Prevention

*Controlled conditions for production process—controlled conditions are not maintained for the implementation of actions to prevent human error.*

### Containment Action

1. Identify Critical Areas Prone to Human Error:
   - Review ongoing processes and tasks to pinpoint areas where human errors are likely to occur or have occurred.
   - Focus on high-risk areas impacting product quality, safety, or compliance.
2. Provide Immediate Oversight:
   - Assign experienced personnel to oversee tasks and monitor for potential errors in real time.
   - Implement a double-check system for critical processes to catch and correct errors before they propagate.
3. Enhance Communication and Clarity:
   - Provide clear, immediate instructions for affected processes.
   - Use temporary visual aids, labels, or checklists to minimize confusion.
4. Suspend High-Risk Tasks If Necessary: Pause processes where human error could lead to significant safety, quality, or operational risks until controls are reinstated.

### RCA

1. Evaluate Process Design: Determine whether processes are overly complex, ambiguous, or lack sufficient safeguards to prevent human error.
2. Assess Training and Competency:
   - Investigate whether employees have received adequate training for their roles and tasks.
   - Identify any knowledge or skill gaps that could lead to errors.
3. Analyze work environment:
   - Review whether environmental factors (e.g., noise, lighting, distractions, ergonomics) contribute to errors.
   - Check whether fatigue or stress levels among employees are impacting their performance.
4. Examine Procedures and Instructions: Assess whether procedures, work instructions, or checklists are unclear, outdated, or incomplete.
5. Review Previous Incidents: Look for patterns in past errors to identify systemic issues contributing to human error.

### Corrective Action

1. Redesign Processes to Minimize Human Error:
   - Simplify and standardize workflows to reduce complexity and ambiguity.
   - Introduce automation where feasible to reduce reliance on manual tasks.
2. Enhance Training Programs:
   - Develop comprehensive training modules to ensure employees are fully competent in their tasks.
   - Include scenario-based training to prepare employees for unexpected situations.
3. Implement Error-Proofing Mechanisms (Poka-Yoke):
   - Use physical or procedural safeguards to prevent errors, such as fixtures or jigs to ensure correct assembly, and automated alerts for process deviations.

4. Improve Procedures and Documentation:
   - Update work instructions, SOPs, and checklists to ensure they are clear, concise, and accurate.
   - Include visual aids, such as diagrams and color-coded systems, to enhance understanding.
5. Enhance Workplace Ergonomics and Environment:
   - Optimize the physical environment to reduce distractions, improve lighting, and ensure comfortable workstations.
   - Implement measures to mitigate fatigue, such as regular breaks and workload balancing.
6. Foster a Culture of Accountability and Feedback:
   - Encourage employees to report errors or near misses without fear of repercussions.
   - Use feedback to improve processes and provide recognition for identifying potential risks.
7. Introduce Monitoring and Performance Tools:
   - Use KPIs to track human error trends and identify high-risk areas.
   - Implement real-time monitoring systems to detect and correct errors promptly.
8. Regular Audits and Reviews:
   - Conduct periodic audits to ensure error prevention measures are effective and adhered to.
   - Use findings to drive continual improvement and reinforce training.
9. Management Commitment and Support:
   - Ensure leadership actively supports and invests in initiatives to prevent human error.
   - Allocate resources for training, automation, and process redesign as necessary.

## CS 77 [AS9100D Clause 8.5.1 h] Uncontrolled Conditions: Gaps in Release and Delivery Processes

*Controlled conditions for production process—controlled conditions are not maintained for the implementation of release, delivery, and post-delivery activities (clause 8.5.1 h).*

### Containment Action

1. Identify Critical Areas Prone to Human Error:
   - Review ongoing processes and tasks to pinpoint areas where human errors are likely to occur or have occurred.
   - Focus on high-risk areas impacting product quality, safety, or compliance.
2. Provide Immediate Oversight:
   - Assign experienced personnel to oversee tasks and monitor for potential errors in real time.
   - Implement a double-check system for critical processes to catch and correct errors before they propagate.
3. Enhance Communication and Clarity:
   - Provide clear, immediate instructions for affected processes.
   - Use temporary visual aids, labels, or checklists to minimize confusion.
4. Suspend High-Risk Tasks If Necessary:
   - Pause processes where human error could lead to significant safety, quality, or operational risks until controls are reinstated.

## RA

1. Evaluate Process Design:
    - Determine whether processes are overly complex, ambiguous, or lack sufficient safeguards to prevent human error.
2. Assess Training and Competency:
    - Investigate whether employees have received adequate training for their roles and tasks.
    - Identify any knowledge or skill gaps that could lead to errors.
3. Analyze Work Environment:
    - Review whether environmental factors (e.g., noise, lighting, distractions, ergonomics) contribute to errors.
    - Check whether fatigue or stress levels among employees are impacting their performance.
4. Examine Procedures and Instructions:
    - Assess whether procedures, work instructions, or checklists are unclear, outdated, or incomplete.
5. Review Previous Incidents:
    - Look for patterns in past errors to identify systemic issues contributing to human error.

## Corrective Action

1. Redesign Processes to Minimize Human Error:
    - Simplify and standardize workflows to reduce complexity and ambiguity.
    - Introduce automation where feasible to reduce reliance on manual tasks.
2. Enhance Training Programs:
    - Develop comprehensive training modules to ensure employees are fully competent in their tasks.
    - Include scenario-based training to prepare employees for unexpected situations.
3. Implement Error-Proofing Mechanisms (Poka-Yoke):
    - Use physical or procedural safeguards to prevent errors, such as fixtures or jigs to ensure correct assembly, and automated alerts for process deviations.
4. Improve Procedures and Documentation:
    - Update work instructions, SOPs, and checklists to ensure they are clear, concise, and accurate.
    - Include visual aids, such as diagrams and color-coded systems, to enhance understanding.
5. Enhance Workplace Ergonomics and Environment:
    - Optimize the physical environment to reduce distractions, improve lighting, and ensure comfortable workstations.
    - Implement measures to mitigate fatigue, such as regular breaks and workload balancing.
6. Foster a Culture of Accountability and Feedback:
    - Encourage employees to report errors or near misses without fear of repercussions.
    - Use feedback to improve processes and provide recognition for identifying potential risks.
7. Introduce Monitoring and Performance Tools:
    - Use KPIs to track human error trends and identify high-risk areas.
    - Implement real-time monitoring systems to detect and correct errors promptly.
8. Regular Audits and Reviews:
    - Conduct periodic audits to ensure error prevention measures are effective and adhered to.
    - Use findings to drive continual improvement and reinforce training.

9. Management Commitment and Support:
   - Ensure leadership actively supports and invests in initiatives to prevent human error.
   - Allocate resources for training, automation, and process redesign as necessary.

## Containment Action

1. Identify and Segregate Affected Products or Services:
   - Immediately review ongoing and completed release, delivery, and post-delivery activities.
   - Quarantine any products or services that may not meet the required standards due to uncontrolled conditions.
2. Implement Temporary Controls:
   - Introduce manual oversight for release and delivery processes to ensure compliance with requirements.
   - Temporarily assign additional personnel to verify delivery and post-delivery activities.
3. Notify Stakeholders:
   - Inform customers, internal teams, and suppliers of any potential delays or issues caused by the lack of controlled conditions.
   - Provide updates on containment measures and timelines for resolution.
4. Verify Affected Outputs:
   - Perform a thorough inspection or testing of products and services to ensure they meet quality and compliance requirements.
   - Replace or recall nonconforming items if they have already been delivered.

## RCA

1. Review Release and Delivery Processes:
   - Determine whether there were clear, documented procedures for release, delivery, and post-delivery activities.
   - Assess whether these procedures were consistently followed.
2. Assess Training and Competency:
   - Investigate whether employees involved in these activities had sufficient training and understanding of controlled conditions.
3. Analyze Monitoring and Verification Systems:
   - Check whether adequate systems were in place to monitor and verify compliance during the release, delivery, and post-delivery stages.
   - Evaluate whether deviations were detected and addressed appropriately.
4. Examine Resource Allocation:
   - Investigate whether inadequate resources (time, personnel, equipment) contributed to uncontrolled conditions.
5. Evaluate Environmental Factors:
   - Assess whether environmental conditions (e.g., storage, transport, handling) were suitable and adhered to during delivery and post-delivery.
6. Identify Communication Gaps:
   - Check whether there was a breakdown in communication between teams, such as unclear instructions or delayed feedback.

### Corrective Action

1. Develop and Enforce Standardized Procedures:
    - Create or update detailed procedures for release, delivery, and post-delivery activities.
    - Include specific requirements for environmental controls, handling, and documentation.
2. Enhance Training Programs:
    - Provide comprehensive training to employees involved in these activities to ensure understanding and adherence to controlled conditions.
    - Conduct periodic refresher courses to reinforce key practices.
3. Strengthen Monitoring and Verification Systems:
    - Introduce automated systems or check-lists to monitor and verify compliance at every stage of release, delivery, and post-delivery.
    - Implement a tracking system to document and audit these activities.
4. Improve Environmental Controls:
    - Establish strict guidelines for storage, transport, and handling to maintain controlled conditions.
    - Use temperature-controlled vehicles or secure packaging for sensitive products.
5. Foster Interdepartmental Communication:
    - Develop clear communication protocols to ensure seamless coordination between teams responsible for release, delivery, and post-delivery activities.
    - Use collaborative tools to track progress and share updates.
6. Implement a Robust Feedback System:
    - Collect feedback from customers and internal teams regarding delivery and post-delivery processes.
    - Use insights to address gaps and improve practices.
7. Conduct Regular Audits and Reviews:
    - Schedule periodic audits of release, delivery, and post-delivery activities to ensure compliance with established procedures.
    - Use audit findings to drive continual improvement.
8. Allocate Sufficient Resources:
    - Ensure adequate personnel, equipment, and budget are available to maintain controlled conditions during these activities.
9. Establish Contingency Plans:
    - Develop contingency plans to address unexpected disruptions during release, delivery, or post-delivery stages.
    - Train employees on how to implement these plans effectively.

## CS 78 [AS9100D Clause 8.5.1 i] Uncontrolled Conditions: Lack of Workmanship Criteria

*Controlled conditions for production process—controlled conditions are not maintained for the establishment of criteria for workmanship (e.g., written standards, representative samples, illustrations) (clause 8.5.1 i).*

### Containment Action

1. Identify and Review Areas Lacking Workmanship Criteria:
   - Audit ongoing processes and tasks to determine where workmanship criteria (e.g., written standards, samples, illustrations) are missing or unclear.
2. Implement Temporary Controls:
   - Assign experienced personnel to oversee tasks and ensure work meets quality expectations based on verbal guidance or existing practices.
   - Use readily available examples or industry standards as interim criteria for workmanship.
3. Inspect Current Outputs:
   - Conduct immediate inspections of completed and in-progress work to identify any deviations or defects.
   - Segregate nonconforming outputs for rework or review.
4. Communicate Expectations:
   - Provide clear verbal or written instructions to employees to ensure consistency and quality in workmanship.
5. Pause High-Risk Processes:
   - Suspend operations for processes where the absence of workmanship criteria could lead to critical quality or safety risks.

### RCA

1. Review Documentation Processes:
   - Investigate whether workmanship criteria were documented adequately during process planning.
   - Assess whether documentation was overlooked, incomplete, or inconsistent.
2. Evaluate Training and Awareness:
   - Determine whether employees and supervisors were adequately trained to understand and apply workmanship standards.
   - Check whether there was confusion or ambiguity in existing guidelines.
3. Analyze Communication Gaps:
   - Investigate whether there were lapses in communication between design, quality, and production teams regarding workmanship expectations.
4. Assess Resource Allocation:
   - Review whether insufficient resources (time, tools, or personnel) contributed to the failure to establish or disseminate workmanship criteria.

5. Examine Change Management Processes:
   - Check whether changes to product designs or processes were implemented without updating workmanship criteria.

### Corrective Action

1. Establish and Document Workmanship Criteria:
   - Develop clear and comprehensive workmanship standards for all processes, including written descriptions, visual illustrations, and representative samples.
   - Use industry benchmarks as a reference where applicable.
2. Standardize Documentation Processes:
   - Create a structured process for documenting workmanship criteria during process planning and updates.
   - Ensure these documents are reviewed and approved by quality and production teams.
3. Enhance Training Programs:
   - Train employees on the importance and application of workmanship criteria.
   - Include practical demonstrations and hands-on sessions using representative samples or illustrations.
4. Improve Communication Protocols:
   - Facilitate regular meetings between design, quality, and production teams to ensure alignment on workmanship criteria.
   - Use collaborative tools to share updated documentation and feedback.
5. Implement Visual Aids and Reference Materials:
   - Display visual aids (e.g., photos, diagrams, examples) at workstations to guide employees in meeting workmanship standards.
   - Create easy-to-access repositories for all workmanship-related documentation.
6. Strengthen Monitoring and Review Systems:
   - Conduct periodic audits of processes to verify adherence to established workmanship criteria.
   - Implement real-time monitoring systems to catch deviations early.
7. Allocate Necessary Resources:
   - Ensure sufficient personnel, tools, and time are allocated to establish, communicate, and enforce workmanship criteria.
8. Integrate Criteria into Quality Control Checks:
   - Include workmanship standards as part of quality inspection protocols at various production stages.
9. Create a Feedback Loop:
   - Establish mechanisms for employees to provide feedback on the clarity and practicality of workmanship criteria.
   - Use this feedback to refine documentation and training.
10. Conduct Regular Updates:
    - Periodically review workmanship standards to ensure they remain relevant to changing designs, materials, and processes.

## CS 79 [AS9100D Clause 8.5.1 j] Uncontrolled Conditions: Lack of Product Accountability in Production

*Controlled conditions for production process—controlled conditions are not maintained for the accountability for all products during production (e.g., parts quantities, split orders, nonconforming product).*

### Containment Action

1. Perform an Immediate Inventory Check:
    - Conduct a full reconciliation of all parts, materials, and products in production.
    - Verify quantities against production plans and identify discrepancies.
2. Quarantine Affected Products:
    - Segregate any products with uncertain accountability or potential NC.
    - Isolate split orders or partially completed items for review.
3. Suspend Production If Necessary:
    - Halt production in areas where the lack of accountability could cause further issues or errors.
4. Enhance Monitoring:
    - Assign personnel to oversee ongoing production activities to ensure accurate tracking of parts, orders, and nonconforming products.
5. Document the Situation:
    - Record all identified discrepancies and actions taken to address them for further analysis.

### RCA

Steps to identify the underlying causes of the issue:

1. **Review Tracking and Documentation Systems:** Evaluate whether production tracking systems (e.g., software, logs, or physical records) are adequate and consistently used.
2. **Examine Material Handling Processes:** Investigate whether there are gaps in procedures for tracking parts and products as they move through production stages.
3. **Assess Employee Training:** Determine whether personnel are adequately trained to follow accountability procedures, including tracking, recording, and reporting.
4. **Check Communication and Coordination:** Review whether communication between departments (e.g., production, quality, and inventory control) is clear and effective.
5. **Analyze Change Management Practices:** Investigate whether split orders, production changes, or handling of nonconforming products are managed and documented properly.
6. **Evaluate Resource Constraints:** Determine whether insufficient personnel, tools, or time contributed to lapses in accountability.

## Corrective Action

1. Implement a Robust Tracking System:
    - Introduce or upgrade software systems to track parts, quantities, split orders, and nonconforming products in real time.
    - Ensure the system provides alerts for discrepancies or missing items.
2. Standardize Accountability Procedures:
    - Develop clear, written procedures for tracking all products during production.
    - Include specific steps for handling split orders and documenting nonconforming products.
3. Enhance Training Programs:
    - Train employees on the importance of accountability and the proper use of tracking systems and procedures.
    - Include practical exercises and regular refreshers to reinforce good practices.
4. Improve Material Handling Processes:
    - Introduce measures like barcoding, RFID tags, or color-coded labels to ensure accurate tracking of parts and products.
    - Establish checkpoints at critical production stages to verify quantities and status.
5. Strengthen Nonconforming Product Controls:
    - Create a dedicated area for quarantining nonconforming products with clear labelling and tracking protocols.
    - Ensure timely documentation and resolution of NC issues.
6. Enhance Communication Protocols:
    - Establish regular communication between production, quality, and inventory teams to ensure alignment on accountability practices.
    - Use collaborative tools to share updates and resolve discrepancies quickly.
7. Conduct Regular Audits and Reviews:
    - Schedule periodic audits to verify compliance with accountability procedures.
    - Use findings to identify and address gaps in the system.
8. Allocate Necessary Resources:
    - Ensure sufficient personnel and tools are available to support accurate tracking and accountability.
9. Introduce Performance Metrics:
    - Track KPIs, such as inventory accuracy, resolution time for discrepancies, and compliance with tracking procedures.
    - Use data to monitor progress and drive continual improvement.
10. Establish a Feedback Loop:
    - Create a system for employees to report issues or suggest improvements to accountability processes.
    - Use this feedback to refine procedures and tools.

## CS 80 [AS9100D Clause 8.5.1 k] Uncontrolled Conditions: Gaps in Critical Item Monitoring

*Controlled conditions for production process—controlled conditions are not maintained for the control and monitoring of identified critical items, including key characteristics, in accordance with established processes.*

### Containment Action

1. Identify and Isolate Critical Items:
   - Review production and inventory records to identify critical items that may have been affected by inadequate control or monitoring.
   - Segregate these items for further inspection and validation.
2. Inspect Current Outputs:
   - Perform an immediate review of ongoing and completed products to ensure compliance with key characteristics and specifications.
   - Flag and rework or reject any nonconforming items.
3. Reinforce Monitoring Procedures:
   - Assign experienced personnel to oversee the control and monitoring of critical items temporarily.
   - Implement a double-check system for critical measurements and processes.
4. Notify Relevant Stakeholders:
   - Inform affected teams (e.g., quality, production, and engineering) and, if necessary, customers about the potential risks.
5. Pause High-Risk Processes:
   - Halt production or processing of critical items if the lack of control and monitoring poses a significant risk to quality or safety.

### RCA

1. **Evaluate Process Documentation:** Investigate whether critical items and key characteristics were clearly defined, documented, and communicated during process planning.
2. **Assess Monitoring Tools and Equipment:** Determine whether the tools and equipment used for monitoring critical items were adequate, calibrated, and functioning correctly.
3. **Review Employee Training and Competency:** Check whether employees responsible for controlling and monitoring critical items were adequately trained on established processes.
4. **Analyze Workflow and Resource Allocation:** Investigate whether insufficient time, personnel, or resources contributed to lapses in control and monitoring.
5. **Examine Communication and Handoff Processes:** Review whether there were breakdowns in communication or handoffs between teams responsible for defining, controlling, and monitoring critical items.
6. **Verify Change Management Practices:** Investigate whether changes to critical item specifications or processes were implemented without updating control and monitoring practices.

### Corrective Action

1. Standardize Control and Monitoring Processes:
   - Develop or update detailed procedures for controlling and monitoring critical items, including clear definitions of key characteristics.
   - Ensure these procedures are accessible and consistently followed across all relevant departments.
2. Enhance Training Programs:
   - Provide comprehensive training to employees on the importance of controlling and monitoring critical items and how to follow established processes.
   - Include practical demonstrations and real-life scenarios to reinforce understanding.
3. Upgrade Monitoring Tools and Systems:
   - Invest in advanced tools and technologies (e.g., automated monitoring systems and sensors) to ensure accurate control and monitoring of critical items.
   - Implement alerts and data logging for deviations from key characteristics.
4. Implement a Robust Change Management System:
   - Ensure any changes to critical item specifications or processes are formally documented, reviewed, and communicated.
   - Include updated control and monitoring requirements in change implementation plans.
5. Establish Verification Checkpoints:
   - Introduce verification steps at critical stages of production to confirm compliance with key characteristics and specifications.
6. Strengthen Communication Protocols:
   - Facilitate regular communication between design, quality, and production teams to ensure alignment on the control and monitoring of critical items.
   - Use collaborative tools to track and share updates on critical items.
7. Conduct Regular Audits and Reviews:
   - Schedule periodic audits to ensure critical items are being controlled and monitored per established processes.
   - Use audit findings to identify gaps and implement improvements.
8. Introduce KPIs:
   - Monitor metrics such as compliance rates with key characteristics, incidents of NC, and time to address monitoring failures.
   - Use KPIs to drive continual improvement.
9. Allocate Sufficient Resources:
   - Ensure adequate personnel, time, and tools are available for the control and monitoring of critical items.
10. Create a Feedback Loop:
    - Establish a mechanism for employees to report challenges or suggest improvements related to controlling and monitoring critical items.
    - Use this feedback to refine processes and training.

## CS 81 [AS9100D Clause 8.5.1 l] Uncontrolled Conditions: Gaps in Measuring Variable Data

*Controlled conditions for production process are not maintained for the determination of methods to measure variable data (e.g., tooling, on-machine probing, inspection equipment).*

### Containment Action

**1.** Identify Affected Processes:
- Audit ongoing production processes to determine where variable data measurement methods are unclear, undefined, or improperly implemented.
- Focus on critical operations involving tooling, on-machine probing, and inspection equipment.

**2.** Perform Manual Verification:
- Assign qualified personnel to manually verify variable measurements for critical processes and outputs.
- Use alternative validated methods temporarily to ensure data accuracy.

**3.** Inspect Completed Outputs:
- Review and test finished products to detect any deviations caused by inadequate measurement methods.
- Isolate and rework or reject any nonconforming products.

**4.** Restrict High-Risk Operations:
- Suspend operations where inadequate measurement methods pose a significant risk to product quality or safety.

**5.** Enhance Oversight:
- Implement immediate oversight for processes involving variable data measurement, with additional checks for compliance.

### RCA

1. **Review Process Documentation:** Check whether variable data measurement methods were clearly defined during process planning. Assess whether documentation included details about tooling, on-machine probing, and inspection equipment.
2. **Examine Training and Competency:** Determine whether employees responsible for measurement methods were adequately trained on how to define and use them effectively.
3. **Analyze Equipment and Tooling Adequacy:** Investigate whether the tooling, on-machine probing, and inspection equipment were appropriate, calibrated, and functioning properly.
4. **Evaluate Process Design and Validation:** Review whether measurement methods were validated during the design phase of processes and whether updates were implemented effectively.
5. **Assess Communication and Change Management:** Identify gaps in communication between design, quality, and production teams regarding the definition and implementation of measurement methods.
6. **Check for Resource Constraints:** Investigate whether insufficient resources (e.g., equipment, time, or personnel) contributed to the lack of controlled conditions.

## Corrective Action

1. Standardize Measurement Methods:
   - Develop clear, documented procedures for determining methods to measure variable data.
   - Specify appropriate tooling, on-machine probing, and inspection equipment for each process.
2. Enhance Process Validation:
   - Validate all measurement methods during process design to ensure they are accurate, reliable, and repeatable.
   - Include validation criteria in the documentation for future reference.
3. Upgrade and Calibrate Equipment:
   - Invest in advanced measurement tools and equipment if necessary.
   - Implement a strict calibration schedule to ensure all tools and equipment are accurate and compliant.
4. Train Employees:
   - Provide comprehensive training on the importance of accurate variable data measurement and the proper use of tooling, probing, and inspection equipment.
   - Include hands-on practice sessions and periodic refresher courses.
5. Strengthen Monitoring and Review Systems:
   - Implement automated systems where possible to monitor variable data measurements in real time.
   - Conduct regular reviews to ensure methods remain effective and aligned with process requirements.
6. Improve Communication Protocols:
   - Foster regular communication between design, quality, and production teams to ensure alignment on measurement methods.
   - Use collaborative tools to share updates and feedback on measurement processes.
7. Establish Feedback and Improvement Loops:
   - Create a mechanism for operators to report challenges or suggest improvements in measurement methods.
   - Use this feedback to refine processes and tools continually.
8. Conduct Periodic Audits:
   - Schedule audits to verify that measurement methods are being consistently applied and adhered to across processes.
   - Use audit findings to identify gaps and implement corrective actions promptly.
9. Allocate Necessary Resources:
   - Ensure sufficient personnel, equipment, and time are available to define and implement effective measurement methods.
10. Introduce KPIs:
    - Track metrics such as measurement accuracy, calibration compliance, and defect rates related to variable data.
    - Use KPIs to monitor the effectiveness of corrective actions and drive continual improvement.

## CS 82 [AS9100D Clause 8.5.1 m] Uncontrolled Conditions: Missing In-Process Inspection Points

*Controlled conditions for production process—controlled conditions are not maintained for the identification of in-process inspection/verification points when adequate verification of conformity cannot be performed at later stages.*

### Containment Action

1. Review Current Processes:
   - Conduct an audit of all production processes to identify areas lacking in-process inspection/verification points.
   - Focus on processes where conformity cannot be verified in later stages (e.g., embedded features, hidden characteristics).
2. Implement Temporary Inspection Checkpoints:
   - Establish provisional in-process inspection points to monitor critical characteristics until permanent controls are in place.
   - Use manual or visual inspection techniques where feasible.
3. Inspect Ongoing and Completed Work:
   - Perform thorough inspections of all ongoing and completed items to identify potential NCs.
   - Isolate and address any nonconforming items immediately.
4. Suspend High-Risk Operations:
   - Pause processes where the absence of in-process inspections poses significant quality or safety risks.
5. Assign Oversight Roles:
   - Deploy experienced personnel to monitor critical processes in real time and document any issues or deviations.

### RCA

1. Review Process Planning Documentation:
   - Check whether in-process inspection points were defined during the initial process design phase.
   - Investigate whether critical characteristics requiring in-process verification were overlooked.
2. Analyze Risk Assessment Practices: Evaluate whether risk assessments during process planning adequately identified areas requiring in-process inspections.
3. Assess Employee Training and Awareness: Determine whether employees responsible for process planning and implementation were trained to identify and establish in-process verification points.
4. Examine Communication and Coordination: Investigate whether communication gaps between design, quality, and production teams contributed to the absence of in-process inspection points.
5. Check for Resource Constraints: Identify whether insufficient time, personnel, or equipment led to the omission of in-process inspection points.

6. Evaluate Change Management Processes: Review whether changes to processes or product designs were implemented without updating inspection requirements.

### Corrective Action

1. Define In-Process Inspection Points:
   - Establish clear criteria for identifying in-process inspection/verification points during process design and planning.
   - Document these points in the process control plan and make them accessible to all relevant personnel.
2. Enhance Process Risk Assessments:
   - Incorporate a detailed risk assessment step during process planning to identify critical characteristics that require in-process verification.
   - Use tools like FMEA to prioritize risks and determine inspection needs.
3. Standardize Inspection Protocols:
   - Develop standardized procedures for performing in-process inspections, including methods, tools, and documentation requirements.
   - Ensure these protocols are integrated into the workflow and adhered to consistently.
4. Upgrade Inspection Tools and Resources:
   - Invest in advanced tools and technologies (e.g., automated sensors, cameras, or probes) for real-time in-process verification.
   - Ensure sufficient resources are allocated to establish and maintain in-process inspection points.
5. Train Employees:
   - Provide training to process planners, quality inspectors, and production personnel on identifying and implementing in-process inspection points.
   - Include practical exercises and case studies to reinforce learning.
6. Improve Communication and Coordination:
   - Foster regular communication between design, quality, and production teams to ensure alignment on in-process inspection requirements.
   - Use collaborative platforms to share updates and feedback on inspection protocols.
7. Conduct Periodic Audits:
   - Schedule regular audits of processes to verify the presence and effectiveness of in-process inspection points.
   - Use findings to identify gaps and refine inspection plans.
8. Establish Feedback Mechanisms:
   - Create a system for employees to report challenges or suggest improvements in in-process inspections.
   - Use this feedback to enhance inspection protocols and tools.
9. Integrate Inspection Points into Quality Control Plans:
   - Ensure that in-process inspection points are clearly documented in the quality control plan and integrated into production workflows.
10. Monitor and Track Performance:
    - Use KPIs to track defect rates, inspection effectiveness, and compliance with inspection protocols.
    - Analyze trends to identify areas for continual improvement.

## CS 83 [AS9100D Clause 8.5.1 n] Uncontrolled Conditions: Lack of Evidence for Production and Inspection

*Controlled conditions for production process—controlled conditions are not maintained for the availability of evidence that all production and inspection/verification operations have been completed as planned, or as otherwise documented and authorized.*

### Containment Action

1. Conduct an Immediate Audit:
   - Review all recent production and inspection/verification records to identify missing or incomplete documentation.
   - Determine which operations lack evidence of completion.
2. Isolate Noncompliant Batches:
   - Identify and segregate any products or batches that lack evidence of completed operations (e.g., inspection records and signoffs).
   - Halt further processing of these batches until proper documentation is available.
3. Verify Operations Against Production Plans:
   - Cross-check completed operations with the production plan to ensure that all steps have been performed.
   - Ensure all inspection or verification points have been documented properly.
4. Implement Temporary Documentation Control:
   - Reinforce a temporary system for ensuring documentation is collected and reviewed at each critical step of production and inspection.
   - Assign responsible individuals to manually ensure that all required evidence is recorded.
5. Increase Oversight During Production and Inspection:
   - Assign additional personnel to oversee the documentation of production and inspection operations in real time.
   - Ensure all inspection results are recorded and verified immediately after completion.

### RCA

1. Review Documentation Processes:
   - Examine the process for recording and storing production and inspection records. Determine whether there is a gap in the system that could cause evidence to be lost or not properly documented.
2. Analyze Process Planning and Design:
   - Investigate whether production and inspection operations were properly planned, including the identification of required documentation for each operation.
   - Check whether there were any missing or unclear steps in the production and inspection planning phase.
3. Evaluate Communication and Coordination:
   - Assess whether there was a lack of communication between teams responsible for production, inspection, and documentation.
   - Investigate whether records were not properly transferred or communicated to the appropriate departments or personnel.
4. Examine Employee Training and Awareness:
   - Review whether employees involved in production and inspection are sufficiently trained in the importance of documenting their operations and ensuring all steps are verified.

- Check whether there are any gaps in knowledge regarding the proper methods for documentation.

5. Check for Resource Constraints:
   - Determine whether there were insufficient resources (e.g., time, personnel, or systems) that led to incomplete documentation or a failure to maintain controlled conditions.
6. Assess Management Oversight and Compliance:
   - Evaluate the effectiveness of management controls, including whether there were checks in place to ensure operations and documentation were completed as planned.
   - Review whether there were any recent changes or a lack of enforcement of procedures that might have contributed to noncompliance.

### Corrective Action

1. Standardize Documentation Procedures:
   - Develop and implement a standardized system for recording all production and inspection/verification operations.
   - Ensure the process includes clear requirements for documenting every step, signoffs, and approval at key stages of production and inspection.
2. Automate or Enhance Documentation Tracking:
   - Invest in digital tools (e.g., manufacturing execution systems or electronic logbooks) that automatically track and store production and inspection records.
   - Implement real-time tracking and verification of completed operations to ensure no step is missed.
3. Introduce Audit Trails and Verification Checks:
   - Implement a system that provides an audit trail for every production and inspection operation, making it easy to trace whether any operation was skipped or improperly documented.
   - Include built-in checkpoints for verifying the completion of all required operations before proceeding.
4. Train Employees:
   - Provide thorough training for all employees on the importance of proper documentation and the specific procedures they must follow to ensure compliance.
   - Include examples and scenarios to reinforce the critical nature of complete records.
5. Improve Communication and Coordination:
   - Foster better communication between production, inspection, and quality control teams to ensure they understand the importance of documentation and work together to maintain complete records.
   - Implement regular meetings or updates to ensure everyone is aligned on the documentation requirements.
6. Implement Stricter Oversight and Monitoring:
   - Establish regular audits to ensure that documentation is being maintained as planned, with compliance checks built into the production and inspection processes.
   - Create feedback loops for identifying and addressing issues with missing or incomplete records immediately.

7. Enforce Accountability for Documentation:
   - Assign clear ownership of documentation responsibilities at each stage of production and inspection.
   - Hold individuals accountable for ensuring all required documentation is completed and properly stored.
8. Conduct Regular Reviews and Continual Improvement:
   - Establish a periodic review process to evaluate the effectiveness of documentation practices.
   - Use feedback from audits and reviews to continually improve the process and ensure ongoing compliance.
9. Update and Reinforce Documentation Policies:
   - Regularly review and update documentation policies to ensure they align with industry standards and internal quality requirements.
   - Communicate any updates to all relevant personnel and incorporate them into training sessions.
10. Track Performance with KPIs:
    - Introduce KPIs to track documentation completeness and compliance.
    - Use these metrics to identify trends, and areas of improvement, and take proactive corrective actions.

## CS 84 [AS9100D Clause 8.5.1 o] Uncontrolled Conditions: Gaps in Foreign Object Control

*Controlled conditions for production process—controlled conditions are not maintained for the provision of the prevention, detection, and removal of foreign objects.*

### Containment Action

1. Conduct an Immediate Inspection:
   - Perform a thorough inspection of all affected areas and processes to identify any foreign objects present in the current production or storage environments.
   - Focus on areas with a higher risk of contamination, such as assembly lines, packing areas, or locations where products are exposed to external elements.
2. Isolate Affected Products:
   - Identify and segregate any products or batches that may have been contaminated by foreign objects during production or inspection.
   - Rework or reject nonconforming products immediately to prevent them from reaching customers.
3. Temporary Increase in Monitoring:
   - Implement additional checks at all critical stages of production and packaging to detect and remove foreign objects before products move to the next stage.
   - Assign dedicated personnel to monitor and manually inspect products for contaminants.

4. Implement Short-Term Cleaning Procedures:
   - Carry out a thorough cleaning of production areas, equipment, and tools to remove any foreign objects that might have accumulated.
   - Establish temporary procedures to ensure that workstations and machinery are cleaned regularly during shifts.
5. Install Temporary Physical Barriers or Safeguards:
   - If necessary, install temporary barriers, covers, or safeguards on machinery or equipment to prevent the introduction of foreign objects into the production process.

### RCA

1. Examine Current Procedures for Foreign Object Control:
   - Review the existing procedures for the prevention, detection, and removal of foreign objects.
   - Determine whether the procedures are clear, comprehensive, and effective in identifying and addressing potential sources of contamination.
2. Evaluate Training and Awareness:
   - Assess whether employees have been adequately trained on the importance of preventing and identifying foreign objects in production processes.
   - Investigate whether there are knowledge gaps related to contamination risks or detection methods.
3. Inspect Equipment and Machinery:
   - Investigate whether equipment and machinery are designed, maintained, and cleaned properly to prevent the accumulation of foreign objects.
   - Identify whether any equipment or parts are prone to breakdowns or failure that might lead to foreign object contamination.
4. Assess Environmental Controls and Workplace Conditions:
   - Review the conditions of the production and storage areas, such as cleanliness, airflow, and access to outside contaminants.
   - Identify whether there are areas that are more prone to contamination, such as open production lines or areas without proper filtration.
5. Investigate Material Handling and Storage Practices:
   - Examine whether materials, tools, or components are being handled and stored properly, which might prevent the introduction of foreign objects.
   - Check whether raw materials, tools, or parts are being stored in areas where contamination is more likely.
6. Review Maintenance and Cleaning Schedules:
   - Assess whether there is a lack of regular maintenance, cleaning, and inspection of equipment and facilities, which could allow foreign objects to accumulate undetected.
7. Identify Issues with Supplier Controls:
   - Investigate whether suppliers' materials, components, or packaging are contributing to foreign object contamination.
   - Check whether any incoming materials are not adequately inspected before being used in production.

### Corrective Action

1. Revise and Strengthen Foreign Object Prevention Procedures:
   - Update and formalize procedures for the prevention, detection, and removal of foreign objects.

- Ensure these procedures include detailed steps for identifying potential sources of contamination, using appropriate detection methods, and removing foreign objects effectively.

2. Implement Automated Detection Systems:
   - Install advanced technology, such as metal detectors, X-ray machines, or vision systems, at key points in the production process to automatically detect foreign objects.
   - Ensure these systems are regularly calibrated and maintained to ensure their effectiveness.
3. Establish Regular Cleaning and Maintenance Schedules:
   - Implement and enforce a strict cleaning and maintenance schedule for all production areas and equipment, with a focus on high-risk areas where foreign objects are more likely to accumulate.
   - Ensure that cleaning materials and methods are appropriate for the type of product being manufactured.
4. Increase Training and Awareness:
   - Provide regular training for all employees on how to prevent, detect, and report foreign objects in production areas.
   - Offer refresher courses to maintain awareness and reinforce the importance of contamination control.
5. Review and Improve Material Handling and Storage Protocols:
   - Improve procedures for handling and storing raw materials, tools, and components to minimize the risk of contamination.
   - Ensure proper segregation of materials and that all components are stored in clean, secure areas that prevent foreign object contamination.
6. Enhance Supplier Quality Control:
   - Work with suppliers to ensure that materials, components, and packaging are free from foreign objects before they arrive at the production facility.
   - Implement incoming inspection protocols for all materials to verify their conformity and cleanliness.
7. Improve Facility Design and Environmental Controls:
   - Assess the design of production and storage areas to ensure they are conducive to cleanliness and contamination prevention.
   - Consider implementing air filtration systems, controlled access points, or other environmental safeguards to reduce the risk of foreign object introduction.
8. Create a Feedback Loop for Continual Improvement:
   - Set up a system for employees to report incidents of foreign object contamination and suggest improvements in the detection and removal processes.
   - Use this feedback to refine procedures and implement corrective actions where needed.
9. Perform Regular Audits and Inspections:
   - Conduct periodic internal audits to assess the effectiveness of foreign object control measures.
   - Review audit results to identify gaps in the system and implement corrective actions.
10. Track Performance with KPIs:
    - Introduce KPIs to monitor contamination incidents, detection system performance, and the effectiveness of preventive measures.
    - Analyze trends to identify areas for improvement and ensure ongoing compliance.

## CS 85 [AS9100D Clause 8.5.1 p] Uncontrolled Conditions: Gaps in Utility and Supply Monitoring

*Controlled conditions for production process—controlled conditions are not maintained for the control and monitoring of utilities and supplies (e.g., water, compressed air, electricity, chemical products) to the extent they affect conformity to product requirements.*

## Containment Action

1. Review Current Utility Usage and Conditions:
    - Perform a thorough assessment of the utility systems (e.g., water, compressed air, electricity, chemical products) to identify any immediate issues affecting product conformity.
    - Verify that critical utilities are being supplied at the correct specifications required for production processes and no impurities are affecting the quality of the products. Use a pH meter to check the acidity; pH is measured on a scale from 0 to 14, where 0 is highly acidic and causes corrosion, 7 is neutral like water, and 14 is highly alkaline.
2. Inspect Production Processes:
    - Identify any product batches or ongoing processes that may have been impacted by utility issues.
    - Isolate affected batches to prevent them from progressing further in the production process until conformity can be ensured.
3. Increase Monitoring and Documentation:
    - Introduce immediate checks at key control points to ensure utilities are being supplied at the proper levels.
    - Document any deviations from the required specifications for future reference and corrective action.
4. Implement Temporary Measures to Stabilize Utility Conditions:
    - If there are discrepancies in utility supply, introduce temporary measures (e.g., backup systems, manual adjustments) to stabilize conditions and ensure compliance with product requirements.
5. Notify Relevant Personnel and Stakeholders:
    - Inform all relevant personnel (e.g., production, quality, maintenance) about the potential impact of utility issues on product conformity and the steps being taken to address the situation.

### RCA

1. Assess Utility Monitoring Systems:
    - Review the current systems and processes for monitoring and controlling utilities.
    - Check whether monitoring equipment is calibrated and functioning properly to detect and report deviations in real time.
2. Evaluate Utility Supply and Distribution Systems:
    - Investigate whether there are issues in the utility supply or distribution infrastructure that could lead to fluctuations or interruptions in service (e.g., pressure changes in compressed air, chemical imbalances, water contamination).
3. Examine Production Process Requirements:
    - Review product specifications and process requirements to ensure that utility conditions (e.g., water temperature, air pressure,

chemical purity) are clearly defined and well understood by relevant departments.

4. Investigate Equipment and System Maintenance:
    - Evaluate whether equipment that controls or relies on utilities (e.g., compressors, filtration systems) is well maintained.
    - Identify whether lapses in maintenance schedules or procedures contributed to utility deviations.
5. Review Training and Competency Levels:
    - Assess whether employees responsible for monitoring and controlling utilities are adequately trained in understanding how utility conditions impact product quality and conformity.
    - Check whether there are gaps in knowledge or awareness that contributed to the oversight.
6. Analyze Historical Data and Trends:
    - Review historical data for utility usage and performance over time to identify patterns or recurring issues that might indicate systemic problems.
    - Look for trends in failures or deviations related to specific utilities and correlate them with product NCs.
7. Investigate Communication and Coordination Between Departments:
    - Examine whether there is effective communication between maintenance, production, and quality teams regarding the importance of maintaining controlled utility conditions for product conformity.

## Corrective Action

1. Establish Robust Utility Monitoring and Control Systems:
    - Implement automated monitoring systems that provide real-time data on the condition of critical utilities (e.g., water temperature, pressure, chemical levels) during production.
    - Set up alarms and alerts to notify personnel of any deviations from the required utility parameters.
2. Define Clear Utility Specifications for Production Processes:
    - Create detailed utility requirement specifications for each product or process to ensure that all critical parameters are well defined and understood by production teams.
    - Ensure these specifications are documented, reviewed regularly, and communicated effectively to all stakeholders.
3. Improve Equipment Maintenance and Calibration Procedures:
    - Develop and enforce a strict maintenance and calibration schedule for equipment that controls or monitors utilities, including backup systems in case of failure.
    - Conduct routine checks to ensure that equipment is functioning properly and is calibrated to meet product requirements.
4. Implement Contingency Plans for Utility Failures:
    - Develop contingency procedures to address situations where utilities cannot be supplied at the required levels.

- Ensure backup systems (e.g., secondary water or air supply, alternative chemicals) are available to maintain product quality in case of utility disruptions.

5. Increase Employee Training and Awareness:
   - Provide comprehensive training for all relevant personnel (e.g., production, maintenance, quality) on the importance of utilities in maintaining product quality and how to monitor and control them.
   - Conduct refresher courses periodically to ensure ongoing awareness and skill retention.
6. Establish Cross-Functional Teams for Utility Management:
   - Create a cross-functional team of production, quality, maintenance, and utility experts to oversee the management of utilities in the production environment.
   - Hold regular meetings to review utility performance, address concerns, and implement improvements as needed.
7. Implement Preventive and Predictive Maintenance Strategies:
   - Shift toward predictive maintenance approaches (e.g., sensor-based monitoring, data analysis) to detect early signs of potential utility-related issues and address them proactively.
   - Ensure that any potential utility disruptions are identified before they affect production.
8. Perform Regular Audits and Inspections:
   - Schedule regular audits of utility usage and control systems to ensure they are functioning as intended and in line with product requirements.
   - Use audit results to identify areas of improvement and implement corrective measures.
9. Review and Strengthen Supplier Relationships:
   - Work closely with utility suppliers (e.g., water, gas, electricity, chemical) to ensure that the quality and reliability of their products meet the needs of the production processes.
   - Establish clear communication channels to quickly resolve any issues that might impact utility conditions.
10. Track Performance with KPIs:
    - Implement KPIs to track utility consumption, performance, and any incidents of deviations that affect product conformity.
    - Regularly review performance data and use it to drive continual improvement in utility management.

## CS 86 [AS9100D Clause 8.5.1 q] Uncontrolled Conditions: Gaps in Product Release and Recall Tracking

*Controlled conditions for production process—controlled conditions are not maintained for the identification and recording of products released for subsequent production use, pending completion of all required measuring and monitoring activities, to allow recall and replacement if it is later found that the product does not meet requirements (clause 8.5.1q).*

### Containment Action

- **Identify and Segregate Affected Products:** Immediately review products released for use before full completion of required measurements and monitoring and segregate them to prevent further use until full verification is completed.
- **Verify Current Status:** Check whether any products released without full testing have already been used in production or are in circulation, and document their status for potential recall.
- **Immediate Recording and Tracking:** Implement manual tracking for any products in progress, ensuring identification and documentation are completed at all stages until full validation is finished.

### RCA

- **Process Review:** Investigate why products were released without completing the necessary measuring and monitoring activities.
- **Examine Documentation Procedures:** Check whether the identification and recording processes for incomplete product verification were unclear or inadequately followed.
- **Assess Training and Awareness:** Determine whether personnel were unaware of the importance of completing all required testing before release.
- **Review Communication and Oversight:** Investigate whether communication gaps or insufficient oversight led to the premature release of products for use.

### Corrective Action

- **Revise Procedures:** Update and enforce procedures ensuring that no products are released for use without completing all required measurements and monitoring activities.
- **Implement Real-Time Tracking Systems:** Introduce automated tracking systems to ensure all incomplete products are properly identified and logged, preventing further use until fully verified.
- **Training and Awareness Programs:** Conduct training sessions for all relevant personnel to reinforce the importance of proper product verification and documentation processes.
- **Introduce a Recall and Replacement Protocol:** Develop a clear, actionable protocol for quickly recalling and replacing nonconforming products if discovered after release.

## CS 87 [AS9100D Clause 8.5.1.1] Certification Audit at AeroTech: Compliance, Improvement, and a Costly Omission

It was a crisp morning at AeroTech Manufacturing, a leading aircraft factory known for its precision and excellence. The factory buzzed with activity as engineers, operators, and technicians worked meticulously on various stages of the aircraft assembly process. Today, however, was different. A team from a CB had arrived to conduct a routine audit for the company's AS9100 certification.

AS9100, the international standard for aerospace QMSs (AQMSs), was vital for AeroTech's continued success in the aerospace industry. Certification audits were an essential part of maintaining compliance and ensuring that processes, equipment, and tools were all up to industry standards. However, this audit would be more challenging than others.

The auditors, led by Clara Martinez, had a reputation for being thorough. They began the inspection with an in-depth review of AeroTech's processes. Clara's team was particularly focused on clause 8.5.1.1 of the AS9100 standard, which deals with the "control of equipment, tools, and software programs" used in production.

The clause required that equipment, tools, and software programs used to automate, control, monitor, or measure production processes be validated before they were released for final production. This validation was crucial to ensure the accuracy, reliability, and effectiveness of these jigs, fixtures, templates, and computer numerically controlled (CNC) programs. In addition, the equipment and tools needed to be maintained according to defined storage requirements, with periodic checks and preservation where necessary.

The auditors spent the morning reviewing the factory floor and examining equipment validation records, maintenance schedules, and software versions. However, as Clara examined the documentation, she began to see cracks in AeroTech's adherence to clause 8.5.1.1.

One of the first things Clara noticed was the lack of validation records for several pieces of production equipment. Machines that automated key parts of the production line, including precision CNC machines and robotic arms used for assembly, had not been validated before being released for full production. This was a serious issue. Without validation, there was no guarantee that these tools were operating correctly and could produce parts to the required specifications. The potential risks were too significant, particularly in the aerospace industry, where even the smallest defect could have catastrophic consequences.

Moving on to the software systems that controlled the automated processes, Clara found further discrepancies. The factory used custom software programs to control everything from material handling to quality inspection. However, there were no records of validation for the software before its deployment, and the factory did not have a clear procedure for periodic software reviews or updates. In addition, there was no assurance that the software was being maintained in a way that would prevent degradation in performance over time. Given that the software controlled many critical aspects of production, this was another serious oversight.

Next, Clara examined the storage areas for equipment and tooling that was not in use.

Some pieces of equipment and tooling were found in open storage areas, exposed to dust and temperature fluctuations. There were no specific storage requirements outlined in the factory's QMS, and no periodic checks were conducted to ensure that these items were being preserved in optimal conditions. Some tools had visible signs of wear, while others showed evidence of being stored improperly, which could affect their performance once put back into service.

Clara made notes of all these findings and brought them to her team's attention. The auditors gathered for a meeting to discuss the NCs they had identified.

Clara took the lead. "AeroTech Manufacturing has a serious issue with clause 8.5.1.1," she began. "Equipment, tools, and software programs are not being validated before they are released into production. In addition, there is no clear system for the maintenance, storage, and preservation of equipment and tooling. This presents significant risks, particularly for the safety and performance of the final product."

The other auditors nodded in agreement. They knew the severity of the situation. In the aerospace industry, compliance with AS9100 is not optional—it is a matter of life and death.

Clara continued, "We need to raise a NC against this. The company must take immediate action to validate its equipment, tools, and software before releasing them for production. They also need to implement storage requirements and periodic checks for equipment and tooling in storage."

With a sense of urgency, Clara drafted the NCR, detailing each issue and specifying the corrective actions required. The report outlined the lack of validation for key equipment and software, inadequate storage practices, and the absence of defined preservation procedures. It was clear that AeroTech needed to take swift corrective action to address these deficiencies or risk losing their certification.

Later that afternoon, Clara and her team presented the findings to AeroTech's management. The factory's Quality Manager, Alan Fisher, was visibly concerned as Clara explained the NC.

"Clara, this is a significant issue for us," Alan said. "We pride ourselves on quality, and I understand how serious these lapses are. But we're committed to resolving this. We'll get a team together and start working on corrective actions immediately."

Alan and his team were instructed to develop an action plan that would address each of the NCs. The first step was to implement a validation process for all production equipment and software programs before they were used in production. This would involve establishing a validation team and creating detailed records for each piece of equipment and software.

Next, they would define proper storage requirements for equipment and tooling, ensuring that everything was stored in optimal conditions to prevent damage and quick and easy retrieval for use. Alan also promised that they would introduce periodic inspections to check the condition of tools and equipment that were not in use.

Clara and her team left the factory that day, satisfied with the commitment shown by AeroTech's management. While the NC was a serious issue, the factory was already taking the necessary steps to rectify it.

As days passed, AeroTech worked diligently to implement the corrective actions. Validation protocols were developed for both equipment and software, and a storage system was put in place to ensure that all tools and equipment were preserved in optimal conditions. Regular maintenance schedules and periodic checks were also established, ensuring that everything was maintained to the highest standards.

A few months later, when the auditors returned for a follow-up audit, they were impressed with the changes AeroTech had made. The factory had addressed all the NCs and had gone above and beyond to improve its processes. The corrective actions were effective, and AeroTech was able to maintain its AS9100 certification, continuing its reputation as a leading supplier of high-quality aerospace components.

It was a close call, but in the end, the audit served as a wake-up call for AeroTech, pushing them to improve their systems and ensure the integrity of their products.

### Containment Actions

- **Immediate Validation of Production Equipment:** AeroTech must halt the use of unvalidated equipment and tools and ensure that any equipment or software still in use undergoes an immediate validation process. Any ongoing production that involves unvalidated equipment must be reviewed for quality, and any affected products should undergo additional inspections.
- **Secure Storage of Equipment:** All tools and equipment must be temporarily relocated to a secure storage area with controlled environmental conditions to prevent further exposure to potential damage.

### RCA

- The factory needed to identify why these issues had occurred. Clara recommended that AeroTech conduct an RCA to determine the underlying reasons for the failure to validate equipment and software. This would involve looking at their internal processes, training programs, and how they managed equipment lifecycle and maintenance.
- It was suspected that the lack of an established procedure for validation and inadequate documentation management contributed to the oversight. Additionally, poor communication between departments handling equipment and maintenance might have caused gaps in the preservation practices.

### Corrective Actions

- **Develop a Validation Protocol:** AeroTech needed to establish a formal validation process for all equipment, tools, and software before they could be used in production. This process should include detailed records for each item, including calibration certificates and validation reports.
- **Create Defined Storage Procedures:** The factory must define and implement proper storage requirements for tools and equipment, especially those not in immediate use. These requirements should address factors such as temperature, humidity, and handling to prevent damage.
- **Periodic Checks and Preservation:** AeroTech should introduce a system for periodic inspections and maintenance of tools and equipment in storage to ensure they remain in optimal condition.
- **Training and Awareness:** All relevant personnel should receive training on the new validation, storage, and maintenance procedures to ensure they understand and adhere to the updated requirements.

## Verification and Certification Recommendation

Clara and her team returned to AeroTech the following week to verify the containment actions and corrective measures. They reviewed the action plan and checked that all steps had been put into motion. AeroTech's management had already started implementing the validation protocols and had temporarily relocated equipment to a designated storage area. The RCA was underway, with the involvement of cross-functional teams, and initial findings pointed to communication gaps and inadequate documentation management as contributing factors.

The auditors were satisfied with the progress made, although they acknowledged that the full effectiveness of the corrective actions would not be clear until a later audit. Clara made her recommendation to the CB: **AeroTech Manufacturing should be granted continued AS9100 certification**, subject to further verification during the surveillance audit in one year.

## Effectiveness Verification during Surveillance Audit

As the audit concluded, Clara emphasized that the effectiveness of the corrective actions would be reviewed during the next surveillance audit. This would include confirming the long-term implementation of validation procedures, the maintenance of equipment and tooling, and the ongoing training of staff. If AeroTech maintained these improvements, they would retain their certification. If the issues persisted or the corrective actions were not sustained, it could lead to a suspension of certification.

AeroTech's management team, led by Alan Fisher, acknowledged the importance of this follow-up. They assured Clara that they were committed to continually improving their systems and processes. They understood that while their initial response had been proactive, the true test would come next year when the auditors would revisit their systems and verify the lasting effectiveness of the corrective actions.

The factory had learned a valuable lesson: compliance with AS9100 was not a one-time achievement but a continual process of improvement. With their corrective actions underway, AeroTech Manufacturing was on the right path to not only retain their certification but also to ensure the highest standards of quality and safety in their aerospace production.

## Special Processes—Story: The Omission That Led to Catastrophe

In an aerospace manufacturing facility, the production of quill shafts for aeroengines was a highly specialized operation. These shafts, machined from high-carbon steel forgings, were integral to transmitting drive from the aeroengine to the accessory gearbox, powering vital systems like the fuel pump. The process was governed by stringent quality control protocols—every step mattered.

One key step in the process was **copper plating**, a protective measure against surface corrosion. However, this plating introduced a hidden risk: **hydrogen embrittlement**. During electroplating, hydrogen ions could penetrate the steel beneath the coating, reducing the material's ductility and making it susceptible to brittle failure under stress.

To mitigate this risk, a heat treatment process called *obrabatha*—or de-embrittlement—was essential. This involved heating the plated shaft to 150°C for an hour, driving out the hydrogen ions and restoring the material's integrity.

## The Process Deficiency

The production of these quill shafts relied on a master process sheet, originally written in Russian. During its English translation, a critical term—*obrabatha*—was omitted. As a result, the de-embrittlement operation was entirely overlooked in the manufacturing workflow.

This omission went unnoticed during inspections, as the adverse effects of hydrogen embrittlement could not be visually verified or detected in standard tests. The quill shafts were approved and delivered without anyone realizing the catastrophic consequences waiting to unfold.

## The Catastrophic Failure

Approximately 150 flight hours after installation in a MiG aircraft, one of the quill shafts failed. The failure was sudden and catastrophic, leading to a loss of engine power and, tragically, an aircraft accident. Upon investigation, the root cause was traced back to the missing de-embrittlement process.

This error, though seemingly small, highlighted the devastating consequences of failing to execute and verify special processes in manufacturing.

## Lessons Learned

1. **Understand the Gravity of Special Processes:** Special processes are often unverifiable through inspection alone. Their outcomes rely entirely on strict adherence to established procedures. Omitting even one step can have disastrous effects on product quality and safety.
2. **Documentation is Critical:**
   - Ensure all process sheets are accurate, complete, and reviewed thoroughly during translations or updates.
   - Use cross-verification to catch potential errors in translation or process handovers.
3. **Auditors as Allies:** Auditors have the expertise to identify gaps in process control. Their role is not punitive but preventive, helping organizations strengthen their operations to avoid costly errors. Treat their insights as an OFI.
4. **Continual Learning and Training:** Rely on **authorized sources** for learning and implementing special processes. Operators, engineers, and quality control teams must be well trained and informed about every step in the workflow.
5. **Proactive Quality Management:** Implement robust controls such as:
   - Regular audits of special processes.
   - Checklist-based reviews to ensure every step is performed.
   - Automated alerts or process flags for critical operations.

## Conclusion

This story underscores the importance of vigilance in executing special processes. It serves as a cautionary tale for all involved in high-stakes manufacturing: every detail matters, every step counts, and every error carries potential consequences far beyond the production floor.

When auditors arrive, view their scrutiny not as a challenge but as a safeguard against oversights that could lead to disaster. Let their expertise guide you in refining processes, ensuring every operation is carried out with precision and, ultimately, upholding the safety and reliability of your products.

## CS 88 [AS9100D Clause 8.5.1.2] The Certification Audit at Precision Processors—NC 1

Precision Processors Ltd. (PPL) was a critical supplier to the aerospace industry, providing specialized processes such as heat treatment, chemical coatings, and surface finishing for aircraft parts. The organization had built a reputation for delivering high-quality services, but its operations came under scrutiny during a routine AS9100 certification audit.

AS9100 is the aerospace quality management standard that ensures rigorous controls across the supply chain. A team of auditors, led by David Chang, arrived at PPL to assess their adherence to the standard. The audit focused on clause 8.5.1.2, which requires organizations to establish arrangements to validate and control **special processes**.

### Discovery of the NC

Special processes, by definition, are those where the output cannot be fully verified by subsequent inspection or testing. Instead, the assurance of quality depends on the process itself being properly validated and controlled. David and his team began by reviewing the company's special processes.

The audit revealed troubling gaps in PPL's systems. While the company had documented procedures for some special processes, there was no consistent definition of criteria for reviewing and approving **these processes**. Specifically:

1. **Lack of Criteria for Process Validation:** Processes such as chemical etching and heat treatment lacked clear criteria to determine their suitability. No standardized benchmarks were used to validate whether the processes consistently produced parts that met the required specifications.
2. **Incomplete Approval Records:** The approval of some special processes was informal, with no documented evidence to show that critical parameters had been reviewed or tested before approval.
3. **Inadequate Training and Qualification of Personnel:** While PPL employed skilled operators, there was no systematic method to certify personnel for specific processes. For example, operators performing chemical coatings were not formally evaluated to confirm their competency.
4. **Missing Follow-Up and Revalidation:** The processes lacked scheduled follow-up validations to ensure they continued to meet criteria over time. Changes in materials, equipment, or environmental conditions were not being reassessed for impact on the process.

David brought these findings to the attention of the audit team and discussed them with PPL's Quality Manager, Sarah Kim. She was visibly concerned. The gaps in process validation posed significant risks—not only to the quality of the parts but also to the safety and integrity of the aircraft they were used in.

"This is a serious issue," David said during the closing meeting. "Without defined criteria for validation and consistent controls, the reliability of your special processes cannot be guaranteed. We'll need to raise an NC against clause 8.5.1.2."

### Containment Actions

PPL quickly acted to contain the issue:

- **Suspension of Unvalidated Processes:** All special processes that lacked validation or approval records were temporarily suspended. PPL identified the affected projects and implemented additional inspections to verify the quality of already processed parts.

- **Temporary Measures for Ongoing Work:** The company assigned additional quality inspectors to monitor ongoing processes while corrective actions were being developed.

## RCA

A team was formed to investigate why these gaps had occurred. Their analysis revealed:

- **Inadequate Process Definition:** The organization had not defined clear criteria for validating and approving special processes.
- **Weak Documentation Practices:** A lack of structured documentation led to informal approvals and inconsistent records.
- **Over-Reliance on Experience:** PPL relied heavily on the expertise of senior operators but did not translate this knowledge into documented procedures and training programs.

## Corrective Actions

To address the root causes, PPL implemented the following corrective actions:

1. **Development of Validation Criteria:**
   - PPL established detailed criteria for each special process, including input parameters (e.g., temperature, pressure, chemical concentration) and output requirements (e.g., hardness, coating thickness).
   - A cross-functional team of engineers, quality personnel, and operators reviewed and approved these criteria.
2. **Formalized Process Approval System:**
   - A new approval process was introduced, requiring documented evidence of review and validation before any special process could be used in production.
   - Each process would now require sign-off from a designated authority within the quality assurance team.
3. **Operator Training and Certification:** PPL launched a training program to certify operators for specific processes. Certification requires operators to demonstrate their competency through tests and observed performance.
4. **Scheduled Follow-Up Validations:** A schedule was created for periodic revalidation of all special processes. Changes in materials, equipment, or environmental conditions would trigger an immediate review and revalidation.
5. **Enhanced Record-Keeping:** A centralized database was implemented to store all process validation and approval records. This system would provide easy access for audits and ensure that all required documentation was consistently maintained.

## Audit Closure and Certification Recommendation

David and his team returned for a follow-up visit to verify the implementation of PPL's corrective actions. They reviewed the newly developed validation criteria, examined the training records for certified operators, and checked the documentation for process approvals. They also observed a revalidation trial for one of the heat treatment processes to ensure compliance with the updated criteria.

The auditors were satisfied with the corrective actions taken by PPL. David commended the team for their prompt response and thorough implementation.

"We've verified the containment and corrective actions," David said. "Your organization has made significant improvements.

I'm recommending continued AS9100 certification. However, the effectiveness of these measures will need to be reviewed during the next surveillance audit."

### Effectiveness Verification During Surveillance Audit

As the certification audit concluded, David emphasized that the long-term effectiveness of the corrective actions would be assessed during the next surveillance audit. This would include:

- Reviewing the adherence to the newly established validation and approval criteria.
- Ensuring that all operators remained certified and that new hires underwent the training program.
- Confirming that the follow-up validation schedule was being followed and documented.

Sarah Kim and her team thanked the auditors for their insights. "This was a wake-up call for us," Sarah admitted. "We understand the importance of robust process validation, not just for certification but for the trust our customers place in us. We'll ensure these improvements become a permanent part of our operations."

### Conclusion

The certification audit at PPL highlighted a critical gap in their QMS but also became an opportunity for growth. By addressing the NC with focused corrective actions, PPL not only retained its AS9100 certification but also strengthened its reputation as a reliable supplier in the aerospace industry. As the next surveillance audit approached, PPL was confident that their enhanced systems would pass with flying colors.

## CS 89 [AS9100D Clause 8.5.1.2] The Certification Audit at Precision Processors—NC 2

Precision Processors Ltd. (PPL), a supplier of special process services such as heat treatment and surface finishing for aircraft parts, was undergoing an AS9100 annual surveillance audit. The audit team, led by David Chang, focused on clause 8.5.1.2, which mandates the establishment of arrangements for validating and controlling special processes, including **the determination of conditions to maintain their approval**.

### Audit Findings

During the audit, the team reviewed PPL's procedures and observed significant gaps in the management of special processes:

1. Undefined Conditions for Approval Maintenance:
   - While PPL had initial approval criteria for processes like chemical coatings and heat treatment, they lacked defined conditions for maintaining these approvals over time.
   - No process existed to monitor or reassess critical factors, such as environmental controls, equipment calibration, or material changes, that could impact the process integrity.
2. Missing Periodic Reviews: There were no scheduled evaluations or performance checks to confirm that the processes continued to meet the required standards after their initial approval.
3. Inconsistent Documentation: Records related to process reviews were incomplete, with no evidence of ongoing compliance monitoring or defined intervals for reassessment.

David raised an **NC** against clause 8.5.1.2, noting the risk of process degradation without clear maintenance conditions.

### Containment Actions

PPL immediately implemented short-term measures:

- **Suspension of At-Risk Processes:** Processes without defined maintenance conditions were suspended pending reassessment.
- **Enhanced Monitoring:** Additional checks were introduced for ongoing production to detect any potential quality issues caused by process variability.

### RCA

PPL's quality team identified the root causes:

- **Lack of Defined Maintenance Procedures:** No formal process existed for determining and monitoring conditions for approval maintenance.
- **Inadequate Oversight:** Responsibility for periodic reviews was unclear, leading to inconsistent implementation.

### Corrective Actions

To address the issues, PPL took the following steps:

1. Development of Maintenance Conditions:
    - Defined specific conditions for maintaining approval, including environmental parameters, equipment calibration schedules, and material quality requirements.
    - Created a checklist for each special process to monitor compliance with these conditions.
2. Implementation of Periodic Reviews:
    - Established a schedule for routine evaluations of all approved processes to ensure they continued to meet requirements.
    - Included reassessment triggers for any changes in equipment, materials, or methods.
3. Improved Documentation Practices:
    - Introduced a centralized system to document review schedules, findings, and corrective actions for each process.
    - Assigned process owners to ensure accountability for maintaining records.
4. Training and Awareness:
    - Conducted training sessions for relevant staff to understand the importance of maintaining process approval conditions and their role in compliance.

### Audit Closure and Certification Recommendation

In a follow-up visit, the auditors verified the containment and corrective actions. They reviewed the newly implemented maintenance conditions, documentation practices, and the schedule for periodic reviews. Satisfied with the improvements, David recommended continued AS9100 certification.

### Effectiveness Verification

David emphasized that the effectiveness of these actions would be evaluated during the next surveillance audit. This would include verifying adherence to the periodic review schedule and confirming that maintenance conditions were consistently applied.

### Conclusion

The audit highlighted critical gaps in PPL's approach to managing special processes but also prompted meaningful improvements. By addressing the NC with structured corrective actions, PPL not only retained its certification but also enhanced the reliability of its operations, ensuring continued customer trust.

## CS 90 [AS9100D Clause 8.5.1.2] Facility Maintenance and Revalidation in Aerospace Manufacturing

Precision Processors Ltd. (PPL), a specialized provider of heat treatment and surface finishing services for aircraft parts, was in the process of undergoing an AS9100 certification audit. The audit, led by seasoned auditor David Chang, was focused on ensuring PPL's compliance with clause 8.5.1.2 of the standard. This clause emphasizes the need for organizations to validate and control special processes, including the approval of facilities and equipment used for these processes.

### Audit Findings: Gaps in Facility and Equipment Revalidation and Approval

As the audit team reviewed PPL's special processes, they uncovered several critical shortcomings related to the approval and qualification of facilities and equipment. These gaps put the integrity of the company's special processes at risk, potentially compromising the quality and safety of aerospace parts. The main findings included:

- Facility Approval Process Lacking: PPL lacked a formal process for approving the facilities where special processes were performed. There were no documented assessments to confirm that the environmental conditions, such as humidity and temperature, were suitable for processes like heat treatment or chemical coating.
- Equipment Qualification Gaps: The equipment used in special processes—such as furnaces and coating tanks—did not have documented qualification records to verify their capability to operate within the specified parameters. Calibration records for critical equipment were inconsistent, with some equipment not being calibrated at the required intervals.
- No Reassessment Criteria: PPL had no system in place for periodic reassessments of facilities and equipment, which would ensure continued compliance with approval standards over time, or after repairs and modifications.

David Chang, the auditor, raised a NC against clause 8.5.1.2, stating, "Without established arrangements for approving facilities and equipment, the integrity of special processes cannot be guaranteed. This poses a risk to the quality of aerospace parts and could compromise safety."

### Follow-Up Actions: Immediate Containment

Upon discovering the gaps in facility and equipment approval, PPL immediately took several containment actions to address the identified risks:

- Halt on Unapproved Processes: PPL suspended operations on equipment that lacked documented approvals and redirected work to equipment that had been validated.
- Environmental Monitoring: Temporary monitoring systems were put in place to measure and control environmental conditions in the facility, ensuring that conditions met the process requirements.

### RCA: Identifying the Issues

PPL's quality team conducted a thorough RCA to understand why the NC occurred. The investigation revealed several key factors contributing to the oversight:

- Process Oversight: The absence of a defined procedure for approving facilities and equipment allowed processes to proceed without proper validation and qualification.

- Inconsistent Records Management: Inadequate documentation practices contributed to gaps in calibration and qualification records, making it difficult to demonstrate compliance.
- Lack of Accountability: Responsibility for approving and maintaining facilities and equipment was not clearly assigned, leading to a lack of ownership and oversight.

### Corrective Actions: Addressing the Root Causes

To rectify these issues and prevent future NCs, PPL implemented a series of corrective actions:

- Facility and Equipment Approval Procedure: PPL developed and documented a formal procedure for approving facilities and equipment before their use in production. This process included criteria for assessing environmental conditions, equipment performance, and process compatibility.
- Qualification and Calibration: A comprehensive qualification of all critical equipment was conducted to ensure compliance with process specifications. The calibration schedule was updated to ensure all equipment was calibrated within the required intervals, with automated reminders in place to prevent future lapses.

### Periodic Reassessments

PPL introduced a system for the periodic reassessment of facilities and equipment. This included requalification after repairs, modifications, or changes in environmental conditions to ensure continued compliance.

### Documentation and Accountability

A centralized system for managing records of facility and equipment approvals, qualifications, and calibrations was implemented.

Clear accountability was established, with the maintenance and quality teams now responsible for ensuring that all approvals and qualifications were up to date and properly documented.

### Training and Awareness

PPL conducted training sessions for its staff on the new procedures for facility and equipment approval, emphasizing the importance of maintaining serviceable facilities and equipment to ensure the quality and safety of aerospace products.

### Audit Closure and Certification Recommendation

After implementing these corrective actions, David and his audit team returned for a follow-up audit to verify the effectiveness of PPL's changes. They reviewed the new procedures, inspected qualification records for critical equipment, and observed the new environmental monitoring system in action. Satisfied with the improvements, David commended PPL for their prompt response and thorough implementation of corrective measures.

"This NC has been effectively addressed," David said. "I am recommending the continuation of your AS9100 certification, provided these measures are maintained. The effectiveness of these actions will be evaluated in the next surveillance audit."

### Effectiveness Verification: Ongoing Commitment to Quality

David emphasized that the next surveillance audit would focus on the following areas to ensure continued compliance:

- Adherence to the newly established facility and equipment approval processes.
- Consistency in the periodic reassessment and calibration of equipment.
- The maintenance of comprehensive documentation for all approvals and qualifications.

### Conclusion: The Importance of Serviceable Facilities and Equipment and Periodically Validating Them

The AS9100 certification audit uncovered significant gaps in PPL's management of facilities and equipment used for special processes. However, by swiftly addressing these issues and implementing robust corrective actions, PPL not only retained its certification but also reinforced its commitment to maintaining serviceable facilities and equipment crucial to the production of aerospace parts.

This story serves as a powerful reminder that in industries where safety and precision are paramount, the approval, qualification, and ongoing maintenance of facilities and equipment must never be overlooked. Regular audits, training, and vigilant adherence to standards are vital to ensuring that aerospace parts meet the highest levels of quality and safety.

## CS 91 [AS9100D Clause 8.5.1.2] Certification Audit at Precision Processors: Personnel Qualification

Precision Processors Ltd. (PPL), a trusted supplier of special process services for aircraft parts, was undergoing an AS9100 certification audit. The audit, led by experienced auditor David Chang, focused on clause 8.5.1.2, which mandates organizations to establish arrangements for the validation and control of special processes, including the **qualification of persons** involved in these processes.

### Audit Findings

During the assessment, the audit team identified significant gaps in PPL's approach to qualifying personnel:

1. **No Formal Qualification Criteria:**
   - Operators performing critical processes like heat treatment and chemical coatings lacked documented qualifications.
   - There was no defined process to ensure personnel were trained and evaluated against specific criteria for competency before being assigned to critical tasks.
2. **Inconsistent Training Records:** Training records were incomplete or missing for several operators. Some staff were performing tasks without evidence of formal training or approval from the quality team.
3. **No Ongoing Competency Assessment:** PPL lacked a system to periodically reassess personnel competency, particularly when new equipment, materials, or processes were introduced.

David raised an **NC** against clause 8.5.1.2, emphasizing the risk posed by unqualified personnel in ensuring consistent and reliable outcomes for special processes.

### Containment Actions

PPL quickly implemented temporary measures:

- **Restricted Process Operators:** Limited critical tasks to experienced operators whose qualifications were verified through past performance.
- **Immediate Training for Others:** Initiated on-the-job training for personnel with unclear or incomplete records.

### RCA

PPL's quality team investigated the issue and identified key causes:

1. **Undefined Qualification Process:** No formalized criteria for qualifying or approving personnel for special processes.
2. **Gaps in Record-Keeping:** Training and competency records were not systematically maintained or updated.
3. **Lack of Ongoing Evaluation:** No program existed to recertify operators after changes in processes or technologies.

### Corrective Actions

To address the root causes, PPL implemented the following corrective actions:

1. Establishing Qualification Criteria:
   - Developed detailed criteria for qualifying personnel based on industry standards, process requirements, and customer specifications.
   - Included theoretical knowledge and practical skill evaluations in the qualification process.
2. Formal Training Program:
   - Introduced a structured training program, covering process-specific requirements, equipment operation, and safety standards.
   - Training sessions included assessments to validate competency before operators were authorized for independent work.
3. Ongoing Competency Assessments:
   - Implemented periodic reassessments of personnel to ensure continued competency, especially after process changes or new equipment installations.
4. Centralized Record-Keeping:
   - Created a centralized database to document training, qualification, and reassessment records for all personnel.
   - Established clear timelines for renewing certifications and updating records.
5. Defined Responsibilities:
   - Assigned a dedicated training coordinator to oversee personnel qualifications and ensure compliance with defined criteria.

### Audit Closure and Certification Recommendation

The audit team conducted a follow-up visit to verify the containment and corrective actions. They reviewed the newly established qualification criteria, examined training records, and observed competency assessments in progress. Satisfied with the thoroughness of the response, David commended PPL for their swift improvements.

"I am confident that the risks identified during the initial audit have been mitigated," David stated. "I recommend continuation of AS9100 certification, with the effectiveness of these measures to be evaluated during the next surveillance audit."

### Effectiveness Verification

The next surveillance audit will focus on:

- Ensuring all personnel continue to meet qualification criteria.
- Reviewing the implementation of the periodic reassessment program.
- Verifying that records are consistently maintained and updated.

### Conclusion

The certification audit highlighted a critical gap in PPL's approach to personnel qualification, posing risks to the quality of special processes. However, the organization's prompt actions not only resolved the NC but also strengthened its QMS, ensuring continued compliance with AS9100 and trust in its aerospace services.

## CS 92 [AS9100D Clause 8.5.1.2] Certification Audit at Precision Processors: Method Implementation and Monitoring

During an AS9100 certification audit at Precision Processors Ltd. (PPL), an organization providing special process services for aircraft parts, the audit team, led by David Chang, identified an NC against clause 8.5.1.2. The clause mandates the **use of specific methods and procedures to implement and monitor special processes**.

### Audit Findings

1. **Lack of Standardized Methods:** Procedures for key processes like chemical coating and heat treatment were generic and lacked specific instructions tailored to individual customer or regulatory requirements.
2. **Inadequate Process Monitoring:** No defined methods to monitor critical parameters, such as temperature, dwell time, or chemical concentration, were evident during production.
3. **Inconsistent Records:** Process monitoring logs were incomplete, making it impossible to trace whether parameters were within acceptable limits.

### Containment Actions

- Suspended production on processes lacking clear monitoring methods.
- Temporarily assigned experienced supervisors to oversee critical operations and ensure compliance with customer requirements.

### Corrective Actions

1. **Standardized Procedures:** Developed detailed work instructions for all special processes, specifying step-by-step methods and monitoring requirements.
2. **Process Monitoring:** Installed monitoring systems to capture real-time data on critical parameters, ensuring compliance.
3. **Training:** Trained staff on updated procedures and the importance of monitoring for quality assurance.

The audit team verified the corrective actions and recommended certification, with effectiveness to be evaluated in the next surveillance audit.

## CS 93 [AS9100D Clause 8.5.1.2] Certification Audit at Precision Processors: Record Retention Requirements

Precision Processors Ltd. (PPL), a supplier of special process services for aircraft parts, underwent an AS9100 certification audit. The audit team, led by David Chang, identified an **NC** against clause 8.5.1.2, which requires organizations to establish arrangements for the validation and control of special processes, including **requirements for documented information to be retained**.

### Audit Findings

1. **Incomplete Record Retention Policy:**
    - PPL had no formal policy specifying the type of documented information to be retained for special processes, such as process parameters, approvals, or operator qualifications.
2. **Missing or Inconsistent Records:**
    - Critical records, including heat treatment parameters and chemical concentration logs, were either incomplete or missing for several batches.
    - There was no centralized system to track and store records, leading to risks in traceability and compliance.

### Containment Actions

- Introduced temporary logging procedures to ensure all critical parameters for ongoing batches were documented.
- Assigned supervisors to review and verify logs before the release of parts.

### Corrective Actions

1. **Retention Policy:** Developed a retention policy specifying the records to be maintained, including process parameters, operator qualifications, and monitoring data.
2. **Centralized Documentation System:** Implemented a digital record-keeping system to store, retrieve, and secure process-related documentation.
3. **Training:** Trained operators and supervisors on the importance of accurate and consistent documentation for compliance and traceability.

The auditors verified the corrective actions and recommended certification. Effectiveness will be reviewed during the next surveillance audit to ensure compliance is sustained.

## CS 94 [AS9100D Clause Non-Specific] The Power of Perseverance and AQMS in Aerospace

This story illustrates a valuable lesson in the aerospace industry about the importance of adhering to quality standards and the role of perseverance in achieving success.

A medium-sized enterprise in Hyderabad was engaged in manufacturing sheet metal and machined parts for the automotive sector. When a friend of the MD, who was visiting from abroad, toured the facility, he gave a piece of crucial advice: "Original equipment manufacturers (OEMs) and multinationals do not come often, but when they do, they offer a test by fire—the first article inspection report (FAIR). If your products pass, you earn their trust and a lasting relationship. An effective AQMS is vital to achieving this success."

Motivated by the opportunity, the company aggressively marketed itself and pursued a sizable aerospace contract worth Rs 2000 million. To secure the deal, the customer required the company to implement AQMS and achieve certification. After considerable effort, the company obtained the certification and received an order to produce structural assemblies for FAI.

However, in the rush to meet deadlines and secure the large order, the team neglected to follow certain critical AQMS process effectiveness requirements. Once the first batch of parts was ready for delivery, the customer came for an audit to verify the FAIR. What followed was a painful experience: a three-day audit revealed 23 NCs, including several major ones. As a result, the order was suspended, causing a major setback and embarrassment for the company.

The customer, however, did not walk away. Instead, they allowed the company to rectify the issues through corrective actions. The team worked diligently to address all the identified problems, ensuring they aligned with AQMS standards. When the customer returned for a re-audit, all corrective actions were accepted, and the suspension was lifted.

This experience was a powerful reminder of the importance of QMSs and continual improvement. The MD reflected, "We may not get it right the first time, but we learn from our mistakes. The key is to plan, do, check, and act. The second time, and every time after that, we'll do it right."

By embracing this philosophy, the company not only regained the customer's trust but also set itself up for future success in the aerospace industry, turning initial embarrassment into a lesson in resilience and process excellence.

## CS 95 [AS9100D Clause 8.5.1.3] Certification Audit at Aero Precision: The Missing Check That Changed Everything

At Aero Precision Components Ltd., the team was confident in their manufacturing processes as they welcomed auditor Lisa Hart for an AS9100 certification audit. The company prided itself on producing critical aerospace parts with precision, but Lisa's eagle eyes were on a mission: to ensure compliance with clause 8.5.1.3, which mandates PPV to guarantee product requirements were consistently met.

### The Goal and the Challenge

Lisa's review led her to a precision-machined landing gear component, an integral part of any aircraft. The goal was clear: to demonstrate that their processes could reliably produce this component to exact specifications.

However, two significant issues emerged:

1. Incomplete FAI/PPV Records: The FAI/PPV records for the component failed to include validation of the tooling used in its production—a key step to ensure the tools were capable of meeting tight tolerances.
2. Dimensional Discrepancies in Production: The first production part of the component did not meet specifications. The discrepancy was traced back to an unverified tool setup, revealing that the tools had not been validated before production began.

Lisa issued an NC, noting that without robust FAI/PPV practices, the company could not guarantee consistent quality or product safety.

### Follow-Up Actions: Containment and Correction

The team acted swiftly to contain the issue and resolve the NC:

#### Containment Actions

- Production of affected components was halted until all tooling validation activities were completed.
- All recent FAIs were re-inspected for similar omissions.

#### Corrective Actions: FAI/PPV Process Overhaul

- FAI procedures were revised to make tooling verification mandatory as part of the FAI/PPV process.
- Tooling Validation and Documentation: Critical tools were validated, and records were updated to ensure future traceability.
- Training for Awareness: Operators and quality personnel were trained on the updated FAI process to reinforce the importance of tooling verification.

#### Verification and Resolution

During a follow-up audit, Lisa scrutinized the updated FAI records and observed a live inspection. She confirmed that the corrective actions were effectively implemented and verified that tooling validation was now an integral part of the PPV process.

Satisfied with the improvements, Lisa recommended maintaining the company's AS9100 certification, with a note that the changes' effectiveness would be evaluated during the next surveillance audit.

### Moral of the Story: Checks and Balances Save the Day

This incident emphasized the critical role of robust PPV practices, especially in industries like aerospace where quality and safety are non-negotiable. A simple oversight in tooling validation nearly derailed the company's certification efforts and risked compromising product integrity.

Through swift action, team training, and process enhancements, Aero Precision Components Ltd. not only addressed the NC but also fortified its operations for the future.

Reflection: Never underestimate the power of a small, missing step. In manufacturing, precision is only as strong as the weakest link—and sometimes, that link is a forgotten check. The incident highlighted the importance of robust PPV practices in ensuring product integrity and aerospace safety.

## CS 96 [AS9100D Clause 8.5.2] Stamps Gone Rogue: A Traceability Tale

The certification audit at AeroCraft Components Ltd. was in full swing, led by experienced yet affable auditor Peter Grant. As Peter reviewed clause 8.5.2 concerning **acceptance authority media (AAM)**, his attention was drawn to the handling of approval stamps during a visit to the inspection area.

### Audit Findings

Peter found a few alarming issues:

1. **Misuse of Stamps:** Stamps meant for product approval were left unattended, allowing unauthorized access.
2. **Worn-Out Stamps:** Some stamps were so worn that their markings were barely legible, creating risks to traceability.
3. **Stamp Sharing:** One inspector admitted to lending his stamp to a colleague to expedite approvals.

In a light-hearted moment to ease tension, Peter quipped, "It seems like your stamps have more freedom than your employees!" The team chuckled nervously, but the gravity of the issue was not lost.

The inspector in question, visibly concerned, pleaded with Peter: "Please do not escalate this to management. I assure you we'll fix it immediately!" Peter maintained his professional demeanor, reminding them that AAM was critical for ensuring product integrity.

### Containment Actions

- All unattended stamps were immediately secured in a locked cabinet.
- The shared stamp was confiscated, and alternative approval methods were implemented temporarily.

### Corrective Actions

- **Policy for AAM:** A comprehensive policy was introduced, defining the use, storage, and maintenance of stamps as AAM.
- **Stamp Maintenance:** Worn-out stamps were replaced, and periodic checks for legibility were mandated.
- **Individual Accountability:** Inspectors were trained on the importance of AAM, with each inspector required to sign a declaration of responsibility for their stamp.

### Verification and Closure

During the follow-up review, Peter verified the corrective actions. Stamps were securely stored, replacements were in use, and inspectors demonstrated awareness of their responsibilities. The inspector who had shared his stamp proudly stated, "I now understand the significance of my stamp. It is an AAM, and I'll never share it again. I'll even return it when I go on leave!"

Peter was satisfied and recommended certification but left a parting warning: "I'll return next year to ensure this system remains effective."

The audit ended on a positive note, with AeroCraft Components Ltd. recognizing the value of robust AAM controls in safeguarding quality and compliance.

## CS 97 [AS9100D Clause 8.5.2] Tag It or Lose It: The Traceability Tightrope

During a thorough certification audit at AeroTech Special Processes Ltd., auditor Claire Harrison uncovered significant issues concerning **traceability requirements** under **AS9100, clause 8.5.2**.

### Audit Findings

Claire was reviewing the traceability system for an aircraft delivered to a customer and noticed several gaps:

1. **Identification of Products Throughout Their Life:** She found that identification tags on some components were either incomplete or had worn off during the production process, making it difficult to trace them back to the original aircraft.
2. **Traceability of Products Manufactured from the Same Batch of Raw Material:** There was no clear documentation linking certain components to their raw material batches. Without this, it would be nearly impossible to trace any potential issues back to the source material in case of a defect.
3. **Assembly Traceability:** For a particular assembly, Claire could not trace the individual components back to the assembly process or to the next higher-level assembly, violating the principle of traceability for complex products.
4. **Sequential Records of Production:** There was no comprehensive and easily retrievable sequential record of production, including manufacturing, assembly, and inspection/verification activities for the parts in question.

### NC Issued

Claire issued an **NC** under **clause 8.5.2**, noting that the organization had **not fully met traceability requirements** for its products, particularly in terms of:

- Product identification throughout its lifecycle.
- Traceability from raw material batches to finished goods.
- Traceability of components through various assembly stages.
- The sequential record of product production, assembly, and verification.

### Follow-Up Actions

The CEO of AeroTech quickly responded, recognizing the seriousness of the issue and prioritizing corrective measures:

1. Containment Actions:
    - A review of all affected products was initiated. Any incomplete or worn-off identification tags were updated, and affected parts were marked for traceability review.
    - The raw material batch records were cross-checked, and affected parts were removed from the customer-facing product line.
2. Corrective Actions:
    - **Traceability System Overhaul:** A robust traceability system was introduced, with unique part numbers for each component and assembly. Each part now had clear and durable identification tags that were linked directly to its raw material batch and production history.
    - **Process Improvement:** Procedures were put in place to ensure that every assembly could be traced back to its components, and each stage of production (manufacturing,

assembly, inspection, verification) was recorded sequentially.

- **Training:** Staff received training on the importance of traceability and the new processes for tracking parts and assemblies.

### Verification by the Auditor

Upon returning for the follow-up audit, Claire was pleased to see the improvements:

- **Traceability Records:** The new traceability system was fully implemented, with each part clearly identified and easily traceable to its raw material batch and assembly.
- **Sequential Production Records:** Claire reviewed the sequential production records, which were now organized and easily retrievable, covering all stages from manufacturing to inspection and verification.
- **Assembly Traceability:** She found that every assembly now had clear documentation linking it to its individual components and the next higher-level assembly.

Satisfied with the corrective actions, Claire recommended certification, noting that the changes were effective. However, she reminded the team that she would **verify the effectiveness of the system** during the next surveillance audit to ensure continued compliance with traceability requirements.

### Conclusion

AeroTech Special Processes Ltd. successfully addressed the traceability issues, ensuring that their processes now fully met **AS9100, clause 8.5.2**. The CEO's quick and decisive actions ensured that the company was on track for certification, with a commitment to maintaining the system in the future.

## CS 98 [AS9100D Clause 8.5.3] The Forgotten Tools: Safeguarding Others' Property

During a certification audit at Precision Process Services Ltd., auditor Rachel Adams discovered an **NC** against **AS9100 clause 8.5.3**. A special tool and measuring equipment, borrowed from a customer, were found neglected on the shop floor, mixed with dirt and waste. The items, considered **customer property**, were neither protected nor safeguarded.

### Audit Findings

1. **Neglect of Customer Property:** The special tool and measuring equipment were left exposed, risking damage.
2. **Lack of Process:** The organization did not have procedures to ensure the proper handling, storage, and maintenance of customer property.

Rachel issued an NC to address the lapse and stressed the importance of protecting customer property to maintain compliance and trust.

### Containment Actions

- The neglected tools were immediately cleaned, inspected, and securely stored.
- The customer was informed of the incident and reassured of corrective measures.

### Corrective Actions

1. **Procedure Development:** A documented process for handling customer property, including storage, maintenance, and return, was established.
2. **Training:** Employees were trained on the new procedures and the significance of safeguarding customer property.

3. **Monitoring System:** A tracking log was implemented to record the receipt, usage, and condition of all customer-provided items.

### Verification by the Auditor

Rachel reviewed the corrective actions, inspecting the newly implemented procedures and storage system. She verified that:

- The tools were secured and well maintained.
- Staff demonstrated awareness of the updated policy.

Satisfied with the resolution, Rachel closed the NC and recommended certification. She emphasized that the effectiveness of these measures would be re-evaluated in the next surveillance audit.

## CS 99 [AS9100D Clause 8.5.4] Foreign Object Debris, Forgotten Dates, and the Flight to Safety

The certification audit at AeroTech Special Processes Ltd. was progressing smoothly until auditor Tom Mitchell, known for his engaging and educational audits, made an interesting observation. As he walked through the stores and assembly area, he noticed several key issues related to **FOD** and **shelf-life control** for consumables.

### Audit Findings

1. **FOD in Stores and Assembly Areas:** Tom observed multiple instances of FOD in the stores and assembly areas, including small tools, dust, and packaging materials scattered in high-traffic areas. He shared a cautionary tale from his earlier experiences, recounting an incident where a small piece of debris was left in an engine component, leading to a catastrophic failure during a test flight. "Imagine the consequences of that debris making its way into a high-precision aircraft part," he remarked. His point hit home with the team.
2. **Expired Consumables in Stores:** Tom inspected a section of consumables stored in the warehouse and found several items, including lubricants and sealants, with expired shelf-lives. He mentioned a similar incident where the expired adhesive was used in critical aircraft assembly, leading to delays and safety concerns. "Shelf-life is not just a number on a label; it is a matter of ensuring reliability and safety," Tom said with a serious tone.

## NC Issued

Based on his observations, Tom issued an **NC** against **AS9100, clause 8.5.4**, stating that the organization had not implemented effective provisions for:

- Prevention, detection, and removal of FOD.
- Shelf-life control and stock rotation.

The presence of FOD and expired consumables in critical areas posed risks to the integrity of the aircraft parts being processed.

## Containment Actions

- The FOD in the stores and assembly area was immediately cleaned and inspected for any impact on parts.
- Expired consumables were segregated and replaced with fresh stock.

## Corrective Actions

1. **FOD Prevention and Detection:** A formalized FOD prevention and removal procedure was developed and implemented. FOD audits and daily checks were introduced in both the stores and assembly areas.
2. **Shelf-Life Control:**
    - The CEO implemented a new **FOD prevention procedure**, which included regular inspections and a dedicated cleaning team. A robust system was introduced for tracking shelf-life, including clear labels on consumables, a first-in first-out (FIFO) stock rotation policy, and regular checks to ensure no expired items were in use.
    - A **shelf-life tracking system** was installed, including labels with expiration dates, a strict FIFO policy, and periodic checks for consumable items.
    - Staff training on FOD control and shelf-life management was rolled out across departments.
3. **Training:** Employees received training on the importance of FOD prevention and the shelf-life management of consumables.

## Verification by the Auditor

Tom returned to verify the corrective actions. He found that:

- The assembly area and stores were clean and free from any FOD.
- A new system was in place to track shelf-life, and consumables were clearly labelled with expiration dates.

Tom was pleased with the changes and the organization's commitment to corrective action. He recommended certification, but with a reminder that effectiveness would be verified during the next surveillance audit. "The journey doesn't end here; it is crucial to keep up the effort," he said, leaving the team motivated and aware of the vital role these processes play in aircraft safety and reliability.

Tom concluded, "AeroTech has shown a strong commitment to quality, and these improvements will ensure the safety and reliability of the parts they supply. But the real test is maintaining these systems in the long run."

## CS 100 [AS9100D Clause 8.5.4 [a to e] Beyond ISO 9001 Compliance: Aerospace Excellence with Sector-Specific Best Practices

The rhythmic hum of machines echoed through the production floor of AeroCraft Systems, a budding division of a medium-sized private industry that had recently dared to step into the high-stakes world of aerospace manufacturing. The goal was ambitious: to achieve AS9100D certification within three months—a timeline that left little room for error.

Ravi Nair, the Production Manager, paced through the factory, his brow furrowed with worry. The directive from corporate had been clear and urgent. Certification meant credibility, contracts, and a shot at real growth. But the challenge lay in the details, and the aerospace standard was not just a slightly modified version of ISO 9001—it was a whole new beast.

The Chief Operating Officer, Iyer, brought decades of experience from the automotive sector. He was well versed in ISO 9001 and had led successful quality management implementations before. Confident in his approach, he gathered the team, determined to get the division certified. However, his lack of familiarity with the nuances of AS9100D left critical gaps in their readiness.

"We'll align our processes with ISO 9001, and that should cover the basics," Iyer had declared in their initial meetings. Ravi had his doubts but respected the COO's experience. The team worked day and night, organizing documentation, training staff, and preparing for the stage 2 audit.

When the day of the audit arrived, the atmosphere was electric with anticipation. The lead auditor, Arjun Menon, a seasoned aerospace veteran, meticulously reviewed every aspect of the production process. His keen eye missed nothing.

The storm hit when Arjun issued an NC under clause 8.5.5. The audit revealed that essential provisions for the preservation of outputs—cleaning, foreign object detection and removal, sensitive product handling, special handling and storage, safety warnings, marking and labeling, and shelf-life control and stock rotation—were either incomplete or absent.

"These requirements are sector-specific," Arjun explained, pointing to the clause. "ISO 9001 preservation practices won't cut it here. Aerospace products demand an extra layer of precision and care. Without that, the certification can't move forward."

The air felt heavy as the audit team left. Iyer's confidence was visibly shaken. Ravi, however, saw an opportunity.

"We can fix this," he said, determination blazing in his eyes. "We have to act fast, but it is doable."

The next three weeks became a whirlwind of action. Ravi and his team immersed themselves in AS9100D, studying every clause, every nuance. They took decisive steps to address each NC item:

- **Cleaning:** Dedicated cleaning stations were installed, with clear work instructions and scheduled cleaning intervals. Specialized cleaning equipment suitable for aerospace components was procured, and staff were trained to handle sensitive parts meticulously.
- **Foreign Object Detection and Removal:** The team implemented strict FOD control

procedures, including tool accountability systems, workstation checks, and routine inspections. Anti-static mats and sealed containers were introduced to minimize contamination risks.

- **Special Handling and Storage:** Temperature- and humidity-controlled storage areas were created for sensitive components. Custom fixtures and padding were added to prevent damage during transport, and handling protocols were formalized with mandatory training. There were no hazardous items in the manufacturing unit.
- **Marking and Labeling:** A comprehensive labeling system was introduced, with clear, standardized markings for safety warnings, handling instructions, and component traceability. Color-coded tags and bilingual safety labels ensured visibility and understanding at every step.
- **Shelf-Life Control and Stock Rotation:** Inventory systems were updated to track part expiration dates, with automatic alerts for approaching shelf-life limits. FIFO practices were enforced, and a dedicated team was assigned to oversee stock rotation.

When Arjun returned for the corrective action verification, the transformation was undeniable. Processes were streamlined, documentation was meticulous, and every preservation requirement was met with care and diligence.

"This," Arjun said, after completing the verification, "is what aerospace readiness looks like."

The initial certification was issued, and AeroCraft Systems officially stepped onto the runway of opportunity. The turbulence had been intense, but with perseverance and teamwork, they had soared above it. And for Ravi, Iyer, and the entire division, the journey to excellence had only just begun.

## CS 101 [AS9100D Clause 8.5.5] Post-Delivery Plot Twist: How NC Drove AeroEdge's Success

It was a crisp Monday morning at AeroEdge Solutions, a small aerospace manufacturing company vying for AS9100D certification. The entire team was on edge as the external auditor from AeroCert Global arrived to conduct the final stage of the audit. Weeks of preparation had gone into ensuring compliance with the stringent requirements of AS9100D. Yet, there was an air of uncertainty.

The lead auditor, Victor Lane, was known for his meticulous attention to detail. After an opening meeting with the Quality Manager, Priya Nair, and the leadership team, the audit began. Everything seemed smooth until Victor reached the section on **post-delivery activities** under clause 8.5.5.

### The NC Discovery

Victor sat in the field service office, flipping through the maintenance records and technical documents related to AeroEdge's flagship product, the AEROVent 3000 turbine assembly.

"So, let's talk about how you manage post-delivery activities," Victor said, his pen poised over his clipboard. Priya nodded confidently.

"We ensure customer support with a dedicated hotline, provide maintenance training, and offer a warranty program. Our field service engineers also handle any off-site repairs."

Victor's eyes narrowed slightly. "And how do you handle in-service data collection and analysis? Performance, reliability trends, lessons learned?"

Priya hesitated. "We… uh… receive customer feedback occasionally, but we haven't formalized a process for that yet. It is something we're working on."

Victor made a note. "What about control and updating of technical documentation for product use, maintenance, and repair?"

"We ensure updates, but the process is informal. Our engineers typically handle it directly when issues arise," Priya replied.

Victor frowned. "And controls for off-site work?"

"Well, the field service engineers follow internal guidelines, but there is not a centralized record-keeping system."

Victor's pen moved again. "Last one: customer support resources for obsolescence and spare parts. How are those managed?"

"We provide spare parts, but we haven't formalized a resource allocation plan for obsolescence management," Priya admitted, looking uncomfortable.

### The NC Issued

The audit continued, but the gaps in clause 8.5.5 left Victor no choice. During the closing meeting, he issued an NC for post-delivery activities. He summarized his findings:

- No formalized process for **collecting and analyzing in-service data** (e.g., performance, reliability, lessons learned).
- No clear system for **controlling, updating, and providing technical documentation** for maintenance and repair.
- **Lack of centralized controls for off-site work**, leading to inconsistencies.
- Gaps in customer support planning, particularly for **training, warranties, and obsolescence management**.

"This is a major NC," Victor concluded, his tone professional but firm. "It needs to be addressed before certification can be granted."

### Turning a Setback into Success

Priya and the team were initially disheartened, but they quickly regrouped. They assembled a task force to address the findings. Over the next month, they focused on:

1. **In-Service Data Collection:** The team developed a structured process to collect and analyze in-service data, implementing a software tool to track performance and reliability metrics.
2. **Technical Documentation:** They revamped the documentation process, assigning a dedicated team to control and update all technical manuals.
3. **Off-Site Work Controls:** SOPs and checklists were created for off-site repairs, along with centralized record-keeping.
4. **Customer Support Resources:** AeroEdge formalized training programs, enhanced warranty policies, and developed an obsolescence management plan.

A re-audit was scheduled, and this time, Victor was impressed. The corrective actions demonstrated not only compliance but also a proactive approach to continual improvement. AeroEdge was awarded its AS9100D certification.

### The Lessons Learned

The NC was not just a setback—it was a pivotal learning moment. Priya realized that certification was not just about passing an audit; it was about embedding quality and customer satisfaction into the DNA of the company. As the team celebrated their certification, Priya raised a toast:

"To challenges that make us better—and to the power of persistence!"

### Corrective Actions and Follow-Up Activities

Priya and the leadership team convened immediately after the audit to chart a path forward. The following corrective actions and follow-up activities were implemented over six weeks:

1. In-Service Data Collection and Analysis:
   - **Root Cause:** Lack of a defined process to gather and analyze customer feedback.
   - **Corrective Action:** The team adopted a cloud-based system to systematically collect in-service data, including performance metrics, reliability reports, and lessons learned. Customer surveys and automated data logs from products were integrated.
   - **Follow-Up:** Regularly scheduled reviews of in-service data were established to identify trends and feed them into the design improvement cycle.
2. Technical Documentation Control:
   - **Root Cause:** Informal updates led to inconsistencies in maintenance and repair documentation.
   - **Corrective Action:** A document control specialist was appointed to oversee all technical documents. Centralized software ensured all updates were version-controlled, with access restricted to authorized personnel.
   - **Follow-Up:** Quarterly audits of documentation were initiated to ensure compliance and consistency.
3. Controls for Off-Site Work:
   - **Root Cause:** Absence of detailed procedures for monitoring and reporting field service activities.
   - **Corrective Action:** SOPs were created, detailing step-by-step protocols for off-site repairs and servicing. Checklists and reporting templates were provided to field service engineers.
   - **Follow-Up:** Supervisors conducted random audits of off-site work to ensure adherence to the SOPs.
4. Customer Support Planning:
   - **Root Caus**e: Reactive approach to customer support for training, warranties, and obsolescence management.
   - Corrective **Action:**
     - Training modules for customers were formalized, including scheduled webinars and hands-on workshops.
     - A dedicated team was assigned to proactively manage warranties and spare part inventories.
     - A risk assessment process for obsolescence was introduced, ensuring the long-term availability of replacement parts.
   - **Follow-Up:** Monthly meetings were held to track customer queries and support resource allocation.

### The Re-Audit

Six weeks later, Victor Lane returned for a follow-up audit. This time, AeroEdge's preparation shone through. Victor reviewed the newly implemented processes and spoke with team members responsible for each corrective action.

"I see significant improvements," Victor remarked. "Your approach to collecting in-service data is robust, and the centralized control of technical documentation is commendable. The off-site work controls and enhanced customer support planning demonstrate a clear commitment to quality."

### Certification Granted

The NC was closed, and AeroEdge was granted its AS9100D certification. During the closing meeting, Victor congratulated the team.

"NCs are opportunities to strengthen your system. Your ability to address these issues promptly and effectively reflects the spirit of AS9100D."

### The Road Ahead

Priya reflected on the journey. "This experience taught us the value of embedding quality in every aspect of our operations—not just for certification but for our customers' trust."

As the team celebrated, Priya raised her glass. "To continual improvement and to the challenges that drive us forward!"

## CS 102 [AS9100D Clause 8.5.5] Addressing a Major NC: Lessons from Post-Delivery Activities

During an AS9100D certification audit at AeroEdge Solutions, the auditor, Victor Lane, uncovered a significant gap in the company's post-delivery processes. While reviewing records, Victor discovered that an engine with a critical deviation had been delivered to a customer.

When questioned, Priya Nair, the Quality Manager, explained, "We replaced the engine promptly, and the customer was satisfied."

Victor responded, "While replacing the engine addressed the immediate issue, there's no evidence of a formal investigation, RCA, or documentation of corrective actions. Without these, there's no assurance this is not systemic."

### Major NC Issued

*Victor issued a major NC under clause 8.5.5, stating:*

*Action taken by the organization, including investigation and reporting, is not evident when an engine was found having deviations noticed after it was delivered to the customer.*

Key issues identified:

1. Lack of RCA for the deviation.
2. No documented corrective actions to prevent recurrence.
3. Failure to formally report the issue and resolution to the customer.

### Corrective Actions

The AeroEdge team acted swiftly:

1. **RCA:**
   - A cross-functional team determined that a calibration error in testing equipment during final assembly led to the deviation.
   - Testing equipment was recalibrated, and inspection protocols were updated.
2. **Deviation Handling Process:** A SOP was created, requiring formal investigation, documentation, and customer notification for all deviations.
3. **Training:** Employees were trained on the new SOP and the importance of documentation and communication.
4. **Process Improvements:**
   - A deviation tracking system was implemented to log and monitor all post-delivery issues.
   - Additional quality checks were introduced to catch deviations pre-delivery.
5. **Customer Communication Protocol:** Formal templates for reporting deviations and resolutions to customers were established.

### Follow-Up Audit and Verification

Six weeks later, Victor conducted a follow-up audit. His findings:

1. **RCA:** The original deviation investigation was comprehensive and well documented.
2. **Deviation Handling:** New cases were logged and handled according to the SOP, with clear evidence of RCA and preventive measures.
3. **Customer Reporting:** The affected customer received a detailed report of the deviation, corrective actions, and preventive measures.

Victor interviewed employees and reviewed training records, finding a strong understanding and compliance with the updated processes.

### NC Closed

Victor closed the NC, stating, "AeroEdge has addressed this NC thoroughly, demonstrating significant improvements in post-delivery activities and deviation management."

With their AS9100D certification secured, Priya reflected, "This experience showed us the importance of embedding quality into every stage, even after delivery. It is not just about compliance—it is about trust."

## CS 103 [AS9100D Clause 8.5.5] Overlooked Obligation: Tackling Obsolescence in Aircraft Electronics

AeroSystems Tech, a leader in designing and manufacturing electronic equipment for aircraft, was undergoing an AS9100D certification audit. The audit had been progressing smoothly, with most processes meeting the rigorous standards, until the auditor, Clara Fields, raised an issue during her review of **clause 8.5.5: post-delivery activities**.

While examining warranty provisions and contractual obligations, Clara made a startling observation.

"Your warranty terms are comprehensive regarding performance, maintenance, and support," she said, "but there's no mention of provisions for the final disposal of the equipment after its useful life. Given the environmental and regulatory implications, this is a significant gap, especially for obsolete electronic components."

The management team, led by Quality Manager Rajiv Patel, was taken aback. "We've never had a customer raise this question," Rajiv admitted.

"That's precisely the issue," Clara replied. "Obsolescence and disposal aren't just the customer's responsibility. As part of your commitment to lifecycle management, this falls under your accountability. I'll have to issue a **major NC**."

### The NC Issued

*Clara documented the NC as follows:*

*Clause 8.5.5: Warranty provisions do not include contractual obligations such as the final disposal of the equipment after its useful life by the customer.*

She elaborated:

1. **Lifecycle Management:** The organization has no strategy for addressing the disposal or recycling of obsolete electronic equipment.
2. **Environmental Compliance:** Failure to provide clear guidelines or support for end-of-life disposal risks noncompliance with environmental regulations.
3. **Customer Support:** No mechanisms exist to assist customers in handling obsolescence effectively.

"This oversight could have significant consequences, both environmentally and reputationally," Clara concluded.

### Corrective Action Plan

AeroSystems Tech agreed with Clara's assessment and immediately began crafting a corrective action plan to address the NC.

1. Policy Update: End-of-Life Management Policy: The organization developed a policy outlining its commitment to lifecycle management, including:
   - Support for the safe disposal, recycling, or repurposing of obsolete equipment.
   - Compliance with international environmental regulations.

2. Contractual Revisions: Warranty provisions were updated to include specific clauses on:
   - **Obsolescence Management:** Providing customers with replacement or upgrade options for obsolete equipment.
   - **Disposal Support:** Offering guidelines and resources for environmentally compliant disposal of electronic components.
3. Process Implementation: A lifecycle management team was established to oversee end-of-life support, including:
   - Maintaining an inventory of obsolete components and recommended replacements.
   - Creating disposal guides tailored to customer needs and regional regulations.
4. Training and Awareness:
   - Employees were trained on the new policy and processes, emphasizing the importance of lifecycle responsibility.
   - Customers were informed of the new warranty provisions and disposal support services.
5. Communication with Stakeholders: Key customers and regulatory bodies were engaged to align the policy with industry best practices and expectations.

## Follow-Up Audit and Verification

Three months later, Clara returned for a follow-up audit to verify the implementation of corrective actions:

1. **Policy and Warranty Revisions:** Clara reviewed the updated warranty documents and lifecycle management policy. They were comprehensive, clearly defining AeroSystems' responsibility for disposal and obsolescence.
2. **Implementation Evidence:** The lifecycle management team demonstrated records of customer inquiries about obsolete equipment and their responses, including detailed disposal guides.
3. **Training Effectiveness:** Employees across relevant departments showed a strong understanding of the new processes.

Clara was impressed. "You've not only addressed the NC but have taken proactive steps to lead the industry in lifecycle responsibility," she said during the closing meeting.

## NC Closed and Certification Recommended

Clara officially closed the NC and recommended AeroSystems Tech for AS9100D certification. She commended the team for their dedication to addressing a complex issue.

"This is a great example of turning a challenge into an opportunity," Clara remarked. "I'll return in six months to verify the long-term effectiveness of these changes, but for now, congratulations."

As the team celebrated their certification, Rajiv reflected, "This experience reminded us that quality extends beyond the product—it is about the entire lifecycle. We're not just responsible for what we create but also for what happens after its journey ends."

## CS 104 [AS9100D Clause 8.5.5] Embracing Excellence Through Every Stage of Delivery

The production floor at AeroCraft Industries buzzed with relentless energy. The goal was clear—achieve AS9100D certification in three months. The challenge, however, lay in the gap between aspiration and reality. Ravi, the Production Manager, knew the road ahead would not be easy, especially when the Chief Operating Officer, Iyer, was navigating with an ISO 9001 compass in an aerospace jungle.

Iyer, with decades of automotive experience, had driven teams to ISO success. But aerospace had its own language, and missing even a small requirement could mean catastrophic consequences for safety and reliability. The team prepped for the audit, unaware of the turbulence awaiting them.

The stage 2 audit commenced, led by Arjun Nair, a veteran auditor who had seen it all. It was not long before he uncovered critical NCs under clause 8.5.5—specifically around post-delivery activities.

Arjun issued an NC, citing gaps in essential aerospace requirements:

1. **Collection and Analysis of In-Service Data:** No system to gather performance feedback or lessons learned from field operations.
2. **Control, Updating, and Provision of Technical Documentation:** Documentation updates for product use, maintenance, and repair were inconsistent.
3. **Controls for External Work:** No defined processes for managing off-site repairs or subcontracted tasks.
4. **Product/Customer Support:** No structured approach for handling customer queries, training, warranties, or obsolescence management.
5. **Post-Delivery Problem Resolution:** No formal system for recalls or on-site servicing when issues surfaced after delivery.

Arjun's verdict was clear: "Without closing these gaps, certification is off the table."

Ravi's mind raced back to his previous role at an automobile company, where a proactive recall had saved the company's reputation. He also remembered a laptop manufacturer that had turned a potential disaster into a customer trust-building opportunity through swift post-delivery action.

Determined to turn things around, Ravi rallied the team. They worked day and night to implement corrective actions:

- **In-Service Data Collection:** The team established a feedback loop with customers, collecting performance metrics, failure reports, and lessons learned. A dedicated analysis team translated data into actionable improvements.
- **Documentation Control:** A digital document management system with strict version control was introduced, ensuring up-to-date technical manuals, repair procedures, and overhaul guidelines were always available.
- **External Work Controls:** Clear procedures for subcontractor management were developed,

including mandatory quality audits, real-time progress tracking, and mobile inspection units for off-site tasks.

- **Customer Support Infrastructure:** A 24/7 customer support center was created, handling queries, technical training, warranty claims, and obsolescence planning. The company even set up a rapid-response team for critical component failures.
- **Post-Delivery Problem Resolution:** A robust problem-solving process was designed, featuring RCA, immediate containment measures, and a structured recall/service program. Inspired by the positive recall cases Ravi had witnessed, the team crafted a response protocol that turned potential setbacks into loyalty-building moments.

When Arjun returned to verify the corrective actions, he was astounded by the transformation.

"This is not just compliance," he said, nodding in approval. "This is a blueprint for aerospace excellence."

The AS9100D certificate was awarded, but for Ravi and the team, the true reward was the confidence that every component leaving their facility was supported long after delivery. They had not just built subassemblies—they had built a legacy of trust and commitment.

## CS 105 [AS9100D Clause 8.5.6] The Route Card Revelation: A Case of Uncontrolled Changes

At AeroSystems Tech, the AS9100D audit uncovered a significant gap when the auditor, Victor Lane, reviewed the planning and production processes. Renowned for his knowledge of global standards, Victor meticulously examined **production process changes** and their management.

While reviewing production route cards and design consultation records, Victor noticed discrepancies.

"These amendments on the route card," he pointed out, "do not have evidence of formal review or approval by design or quality team. This violates **clause 8.5.6**: Changes must be reviewed and controlled to ensure continuing conformity with requirements."

### NC Issued

*Victor issued the NC as follows:*

***Clause 8.5.6:** The organization has not reviewed or controlled changes in the production process to ensure continuing conformity with requirements, as seen from the project slip (PS) and amendments noted in the route card.*

### Key Findings

1. Changes to production processes were implemented without formal review or approval.
2. Lack of traceability for design consultation and quality assurance during these changes.
3. Risk to product conformity due to uncontrolled amendments.

## Containment Actions

The team at AeroSystems Tech immediately initiated containment measures to minimize potential risks:

1. Freeze on Production Changes: All nonessential production changes were suspended until formal reviews were conducted.
2. Immediate Audit of Active Route Cards: A cross-functional team reviewed all current route cards for undocumented or unapproved changes.
3. Enhanced Quality Inspections: Additional inspections were conducted on products affected by unreviewed changes to identify any deviations or defects.

## RCA

A structured RCA was performed using the **5 Whys** methodology:

1. Why was the route card amendment not reviewed?
    - Because there was no mandatory process requiring formal reviews for production changes.
2. Why was the process missing this requirement?
    - Because the existing change management procedure was outdated and lacked clear guidelines.
3. Why was the procedure outdated?
    - Because periodic reviews of the procedure were not conducted.
4. Why were periodic reviews not conducted?
    - Because there was no accountability for maintaining and updating the procedure.
5. Why was accountability missing?
    - Because responsibility for change management was not clearly defined across teams.

## Root Cause

Lack of a robust change control process and unclear accountability for managing production process changes.

## Corrective Actions

1. **Revised Change Management Process:** Developed a new change control procedure, requiring:
    - Formal review and approval of all production changes by design and quality teams.
    - Documentation of review decisions and traceability on route cards.
2. **Defined Roles and Responsibilities:** Designated accountability for change management across planning, production, design, and quality teams.
3. **Retrospective Review:** Conducted a detailed audit of past production changes to identify and rectify any uncontrolled amendments.
4. **Training and Awareness:** Conducted mandatory training for all employees on the new procedure, emphasizing the importance of controlled changes and traceability.
5. **Monitoring and Oversight:**
    - Established a **change review board** to oversee all future production changes.
    - Implemented a system for periodic reviews of the change management procedure.

## Verification and Closure of NC

Victor returned to verify the corrective actions. He reviewed:

1. **Updated Procedures:** The revised change control procedure was well documented and implemented.
2. **Evidence of Compliance:** New route cards showed documented reviews and approvals for all changes.
3. **Containment Effectiveness:** Products impacted by earlier uncontrolled changes had passed additional quality inspections, with no major defects found.
4. **Employee Competence:** Training records and employee interviews demonstrated a strong understanding and adherence to the new processes.

Victor was satisfied and closed the NC, stating, "This organization has addressed the root cause and implemented effective corrective actions to prevent recurrence."

### Effectiveness Verification

A year later, Victor returned for a surveillance audit to verify the long-term effectiveness of the changes. There were no instances of uncontrolled production changes. The change review board and periodic reviews ensured sustained compliance.

Victor concluded, "AeroSystems Tech has demonstrated lasting improvement in change management, setting an excellent example of proactive quality assurance."

## CS 106 [AS9100D Clause 8.5.6] Unclear Authority: Addressing Change Approval in Production

During the same AS9100D audit at AeroSystems Tech, the auditor, Victor Lane, raised another critical issue related to **clause 8.5.6: control of changes**. While examining the organization's change management process, Victor discovered a significant oversight.

He observed that while changes to production processes, equipment, tools, or software programs were being made, there was no formal mechanism to identify or authorize individuals responsible for approving such changes.

"This is a critical gap," Victor stated during the closing meeting. "Without clearly defined authorization, there's a risk of unauthorized changes disrupting production conformity. I will need to issue an **NC**."

### NC Issued

*Victor issued the NC as follows:*

***Clause 8.5.6:*** *Persons authorized to approve production or service provision changes are not identified.*

### Key Findings

1. Lack of documentation or a defined process to identify personnel authorized to approve production or service provision changes.
2. Risk of unauthorized changes to processes, equipment, tools, or software programs.
3. No accountability structure for managing and documenting change approvals.

### Containment Actions

The AeroSystems Tech team implemented immediate measures to mitigate the risk:

1. Temporary Approval Freeze: Suspended all non-critical production changes until formal authorization was established.
2. Emergency Approval Team: Formed a temporary team of senior managers to review and approve urgent changes.
3. Audit of Recent Changes: Conducted an audit of recent production changes to ensure no unauthorized adjustments had compromised product quality.

### RCA

The quality and planning teams conducted an RCA using the **Ishikawa (fishbone) diagram:**

- **Processes:** Lack of a defined procedure for assigning and documenting change approval responsibilities.
- **Training:** Employees were not trained to follow a structured authorization process.
- **Tools:** The change management system did not include a field for identifying authorized approvers.
- **Accountability:** Ambiguity in roles and responsibilities regarding change management.

### Root Cause

No formal system or process to define and communicate authority for approving production and service changes.

### Corrective Actions

1. **Define Authorized Personnel:** Created a list of roles and personnel authorized to approve different types of production and service changes, including:
   - **Processes:** Production managers and quality engineers.
   - **Equipment and Tools:** Maintenance supervisors and design engineers.
   - **Software Programs:** IT and software engineering leads.
2. **Update Procedures:** Revised the change control procedure to include:
   - Criteria for determining authorized approvers.
   - A requirement to document approvals on route cards, change logs, and relevant records.
3. **System Enhancements:** Updated the change management system to:
   - Include a mandatory field for selecting an authorized approver.
   - Generate alerts for unauthorized changes.
4. **Training and Communication:**
   - Trained all relevant personnel on the updated process and their roles in approving changes.
   - Communicated the list of authorized approvers across departments.
5. **Monitoring and Oversight:** Established a periodic audit process to ensure compliance with the updated procedure.

### Verification and Closure of NC

Victor returned for a follow-up audit and verified the corrective actions:

1. **Updated Processes:** Reviewed the revised change control procedure, which clearly defined authorized approvers.
2. **Compliance Evidence:** Verified that route cards and change records included documented approvals from authorized personnel.
3. **System Enhancements:** Tested the updated change management system, confirming that unauthorized changes were flagged.
4. **Training Effectiveness:** Interviewed employees and reviewed training records to confirm understanding and compliance.

Victor closed the NC, stating, "The organization has successfully established a robust framework for managing and authorizing changes, ensuring conformity with requirements."

### Effectiveness Verification

A year later, during a surveillance audit, Victor verified the sustained effectiveness of the corrective actions:

- All production and service changes were approved by designated personnel.
- No unauthorized changes were identified.
- The periodic audit process ensured continued compliance.

Victor commended AeroSystems Tech, saying, "Your commitment to addressing this NC has not only resolved the issue but also strengthened your change management processes."

## CS 107 [AS9100D Clause 8.6] The Race Against Time: An Incomplete Delivery

At AeroSystems Tech, a production team faced mounting pressure to meet an impending delivery date for a critical aircraft assembly. In their rush, they released the assembly to the customer despite final quality checks and certain minor operations being incomplete. This oversight was discovered during a subsequent **surveillance audit** for AS9100 certification by the meticulous auditor, Clara Fields.

### NC Issued

During her review of production and service processes, Clara uncovered documentation revealing the premature delivery. After interviewing staff and examining the production logs, she noted:

> ***Clause 8.6:*** *The release of products to the customer was earlier than the completion of all the required activities (final operations) on the aircraft. Prior approval had not been obtained from the customer or regulatory authority.*

### Key Findings

1. The assembly was released without completing all required operations and final inspections.
2. No documented approval from the customer or regulatory authority was obtained before the release.
3. The risk of product NC was increased due to the incomplete work.

"This lapse undermines the integrity of your production and quality assurance process," Clara stated firmly.

### Containment Actions

The organization took immediate steps to address the issue:

1. **Customer Notification:** Informed the customer about the premature release and provided an update on the completion of pending tasks.
2. **Recall and Rectification:** Arranged for the assembly to be returned or for on-site rectification by a specialized team.
3. **Enhanced Oversight:** Suspended all pending deliveries for verification of completion status.

### RCA

An RCA was conducted using the **5 Whys** method:

1. Why was the assembly released without completing all required activities?
   - The production team prioritized meeting the delivery deadline over process compliance.
2. Why was there a rush to meet the deadline?
   - The delivery schedule was not adjusted to accommodate delays in earlier production stages.
3. Why were the delays not accounted for?
   - No mechanism existed for formally escalating or adjusting deadlines in case of delays.
4. Why was there no escalation mechanism?
   - The organization lacked a robust procedure for managing delayed production timelines.
5. Why was this procedure missing?
   - Production planning processes were not periodically reviewed or updated to address such risks.

### Root Cause

Pressure to meet deadlines without a mechanism to address delays or escalate for deadline adjustments, coupled with inadequate oversight of delivery readiness.

### Corrective Actions

1. **Strengthened Delivery Readiness Procedures:** Updated the delivery readiness checklist to include mandatory verification of the following:
   - Completion of all required operations and inspections.
   - Formal documentation of customer or regulatory authority approvals for exceptions.
2. **Improved Production Planning:** Introduced a formal **escalation process** for delayed production timelines:
   - Teams must notify management of delays early, with options for adjusting deadlines.
   - Delivery deadlines can only be changed with formal approval from senior management.
3. **Employee Training:** Conducted training sessions for production and quality assurance teams on:
   - The importance of compliance with AS9100 requirements.
   - Procedures for handling exceptions, including obtaining necessary approvals.
4. **Monitoring and Accountability:**
   - Assigned responsibility to the quality manager to oversee the readiness of all deliveries.
   - Established periodic audits of delivery processes to ensure ongoing compliance.

### Verification of Corrective Actions

Clara Fields returned a month later to verify the implementation of corrective actions. Her findings included:

1. **Updated Documentation:** The delivery readiness checklist and escalation procedures were clearly defined and implemented.
2. **Compliance Evidence:** Recent delivery records showed proper verification of completion and, where applicable, documented customer approvals.
3. **Employee Understanding:** Interviews with staff confirmed their understanding of the revised procedures.

Clara was satisfied with the corrective actions and recommended the continuation of AS9100 certifications. However, she added a condition:

"I will return next year to verify the long-term effectiveness of these measures," Clara noted.

### Effectiveness Verification

During the following year's surveillance audit, Clara reviewed the delivery records and escalation logs. Her findings included:

- All deliveries were compliant with the updated readiness procedures.
- Delayed production timelines were formally escalated and managed without compromising product quality or requirements.
- No further instances of premature deliveries were observed.

Clara concluded, "AeroSystems Tech has demonstrated sustained improvement and a strong commitment to meeting AS9100 standards. Your actions have ensured product integrity and customer satisfaction."

The experience reinforced the organization's dedication to compliance, quality, and continual improvement.

## CS 108 [AS9100D Clause 8.6] Incomplete Documentation: The Missing Link in Conformity

During a **surveillance audit** at AeroSystems Tech, a company specializing in the design and manufacturing of high-performance aerospace equipment, the auditor, Clara Fields, noticed an issue in the **production and service process**. While reviewing the documentation related to recently delivered aircraft and aero engines, Clara discovered that there was a gap in the documentation that could undermine the evidence of product conformity.

### NC Issued

Clara observed that the organization had delivered several aircraft and engines to customers with **pending work** and **deferred defects** to be completed after delivery at the operating units. However, the relevant **documentation** did not reflect this information or provide adequate evidence that these defects would be corrected promptly. As a result, the product was delivered without clear proof that it fully met all defined requirements.

*Clause 8.6: "The organization shall ensure that retained documented information provides evidence that the products and services meet the defined requirements."*

*Clara issued the following NC:*

### NC

The organization has not fully complied with ensuring that retained documented information provides evidence that the products and services meet the defined requirements, as there are pending works and deferred defects in the aircraft/aeroengine, with no documentary proof to demonstrate how these will be completed after delivery.

## Key Findings

1. **Pending Work and Deferred Defects:** Aircraft and aeroengine deliveries contained unresolved defects or incomplete tasks that were deferred to later stages, but this was not documented appropriately.
2. **Lack of Documentary Evidence:** No clear documentation was available showing the planned actions to address the pending work or deferred defects.
3. **Compliance Risk:** The absence of proper documentation for deferred actions created a risk that the product did not fully conform to the agreed-upon requirements at the time of delivery.

## Containment Actions

The organization quickly implemented the following containment actions:

1. **Customer Notification:** The company notified affected customers about the deferred defects and explained the corrective actions that would be taken.
2. **Temporary Hold on Further Shipments:** All new deliveries were temporarily halted to ensure that any deferred defects or pending work were fully documented and resolved before shipping.
3. **Immediate Review of Delivered Units:** A thorough review was conducted to verify whether similar issues existed with other delivered aircraft or engines, ensuring all outstanding issues were identified and tracked for resolution.

## RCA

A detailed RCA was carried out to determine why the issue of incomplete documentation occurred. Using the **5 Whys** technique, the analysis revealed the following:

1. Why was there missing documentation on pending work or defects?
   - The final delivery documentation was rushed, and certain items were marked for post-delivery correction but were not formally documented.
2. Why was the documentation rushed?
   - Pressure to meet delivery deadlines led the production team to prioritize the shipping of completed units, deferring unresolved issues.
3. Why was there no formal record of deferred defects?
   - The procedure for documenting deferred defects was not clearly defined or integrated into the final delivery process.
4. Why was the procedure not clearly defined?
   - The organization had not previously experienced major issues with deferred work, and as a result, the process had not been reviewed or updated regularly.
5. Why had the process not been reviewed?
   - There was no formal process for periodically assessing and updating the documentation procedures, particularly for deferred work and defects.

## Root Cause

The lack of a formal, consistent process to document and track deferred defects and pending work that would be completed post-delivery.

## Corrective Actions

To resolve the issue and prevent recurrence, AeroSystems Tech implemented the following corrective actions:

1. **Update Documentation Procedures:**
   - **New Documentation Template:** Introduced a **delivery documentation template** that explicitly includes sections for deferred defects and pending work, along with expected completion dates and responsibilities.
   - **Detailed Record of Post-Delivery Actions:** Created a section in the delivery report for documenting any pending work or defects and clearly specifying actions required for post-delivery resolution.
2. **Implement Approval for Deferred Work:**
   - Introduced a formal **approval process** where deferred defects and pending work must be documented and signed off by the customer or regulatory authority before the product is shipped.
   - **Customer Agreement:** In cases where the work is deferred, customers must acknowledge and approve the plan for post-delivery completion.
3. **Enhance Training and Awareness:**
   - Conducted **training sessions** for all production, quality assurance, and documentation teams on the new procedures, ensuring they understood how to document deferred defects and pending work correctly.
   - Focused on the importance of accurate, complete documentation as part of the overall QMS.
4. **Regular Audits of Documentation Compliance:**
   - A new **audit process** was introduced to review and verify that all completed products had documented evidence of conformity, including any deferred defects or post-delivery work.
   - Management implemented periodic checks to ensure that every product delivered met the defined requirements at the time of shipment.

## Verification and Closure of NC

Clara returned for a follow-up audit to verify the implementation of corrective actions.
She reviewed:

1. Updated Documentation: The new templates for documenting deferred work and defects were in place, and records showed that all recent deliveries included this information.
2. Customer Approvals: Documentation included customer approval for deferred work, where applicable.
3. Audit Trail: Internal audits confirmed that the new process for documenting and tracking deferred defects was being followed.
4. Training Effectiveness: Interviews with staff revealed that they understood and adhered to the new documentation procedures.

Clara closed the NC, stating, "AeroSystems Tech has effectively addressed the issue of incomplete documentation. The corrective actions are comprehensive and well implemented."

### Effectiveness Verification

Clara concluded the audit with a recommendation for the continuation of certification, adding a condition:

"I will return in a year to verify the long-term effectiveness of the corrective actions," she noted.

A year later, Clara revisited AeroSystems Tech for the effectiveness verification. During the follow-up audit, she found:

- No instances of incomplete documentation for deferred defects or pending work.
- All deliveries were fully documented, including customer approval for deferred issues.
- The new procedures were consistently followed, and no further gaps in documentation were identified.

Clara officially recommended the **continuation of AS9100 certification**, praising AeroSystems Tech for its commitment to continual improvement and high-quality standards.

## CS 109 [AS9100D Clause 8.6] Audit Finding: A Case Study in Process Improvement and Business Excellence

During a routine audit of Aerofix Solutions, a leading maintenance, repair, and overhaul (MRO) organization in Southeast Asia, the auditor identified a significant NC in the core maintenance process. This case demonstrates how such findings can catalyze meaningful improvements, enhancing both the organization's reputation and its operational effectiveness.

### The Incident

Before delving into the audit, the auditor reviewed Aerofix Solutions' achievements, continual improvement initiatives, and performance metrics from the previous year. It was also noticed that the organization had no NCs reported in any of the previous audits for compliance to AS9100 despite the best efforts by the auditors to detect a NC in the excellent organization. The audit then proceeded to the MRO workshop, where the main landing gear assembly was inspected and assembled after acceptance.

While reviewing the final inspection and acceptance records, the auditor discovered a discrepancy: "Nil, Replaced," "Nil, Outhouse repair," etc., were recorded where numeric measurement results were expected to be filled in. Further investigation into the records of inspection of aircraft serviced in the recent past revealed:

- Measurement results were not documented in numeric terms as required with the ID of the instrument used.

- Comments were written in place of quantifiable data.
- Acceptance criteria were absent, preventing a direct correlation with measurement results.

When questioned, the process owner could not provide a satisfactory explanation for ensuring the required assembly interference without numeric data.

---

*The auditor issued an NC as follows under clause 8.6: "Retained documented information does not include evidence of conformity with the acceptance criteria."*

*Evidence: "Interference of 15 microns was to be maintained while assembling the bearings in the undercarriage as per the maintenance manual. Measurement is recorded only for the inner diameter of the bearing seat, and the outer diameter of the bearing is not recorded. How the interference of 15 microns is ensured is not known."*

---

Recognizing the issue, the auditee acknowledged the NC. By the time the word reached the ears of top management, senior managers personally verified the facts on the spot in the presence of the Indian auditor, ensuring first-hand information was available to address the issue effectively. The CEO inquired with the auditor what to do as the immediate containment action. The auditor explained "Please redo the measurements and record it if the aircraft is not delivered to the customer. Examine inspection records of aircraft serviced in the recent past for corrections. If the NC is delivered, perhaps you must take the recall procedure or send a party to the customer base for the re-examination."

### Immediate Response

Top management, led by CEO Jonathan Miller, swiftly intervened. Senior managers committed to rectifying the issue promptly and expressed their good intentions to improve processes. The proactive approach demonstrated Aerofix Solutions' commitment to enhancing its AQMS.

### Containment

- A halt was placed on similar assembly operations until compliance could be ensured.
- Additional inspections were conducted to verify the accuracy of completed assemblies.
- Staff were briefed on the criticality of numeric documentation.

### RCA

The RCA identified:

1. **Lack of Clear Guidelines:** Procedures for documenting measurement results were ambiguous.
2. **Inadequate Training:** Staff were not adequately trained to adhere to standard recording practices.
3. **Process Oversight:** Insufficient checks during the final inspection phase allowed the deviation to persist.

### Corrective Actions

1. Revised Documentation Procedures:
    - Clear instructions were issued mandating the use of numeric data for measurement results.
    - Templates were updated to include mandatory fields for numeric entries.

2. Training Programs:
   - Conducted workshops on proper documentation practices.
   - Reinforced the importance of traceability and compliance in AQMS.
3. Enhanced Oversight:
   - Implemented a two-tier review system for inspection records.
   - Assigned a dedicated quality assurance team to oversee compliance.
4. Audit Follow-Up:
   - Scheduled a mid-term audit to ensure sustained compliance.
   - Introduced periodic internal audits to preempt potential deviations.

## Adding Value and Demonstrating Good Intentions

The NC identified provided a genuine OFI. By addressing the issue, Aerofix Solutions not only corrected an aberration in practice but also demonstrated transparency and accountability. Admitting the mistake liberated the organization's processes from indiscipline, replacing vague explanations with quantifiable, reliable data. This improved clarity and traceability are vital for sustaining business excellence.

Top management's quick response, including personal involvement in verifying facts and implementing corrective actions, showcased their good intentions and commitment to process integrity. This proactive approach reinforced stakeholder trust and enhanced the organization's reputation within the industry.

## Outcome and Organizational Impact

The proactive response demonstrated Aerofix Solutions' commitment to quality and business excellence. By addressing the NC head-on, the organization:

- Enhanced its processes, ensuring greater accuracy and reliability.
- Strengthened its reputation among stakeholders for transparency and accountability.
- Fostered a culture of continual improvement, turning an audit finding into a growth opportunity.

## Conclusion

The incident at Aerofix Solutions highlights the value of external audits in identifying and addressing process deficiencies. The leadership's willingness to embrace feedback and act decisively not only rectified the immediate issue but also reinforced the organization's position as a trusted player in the MRO industry. By transforming a "whimper" into a whisper of determination, Aerofix Solutions exemplified how embracing NCs can pave the way for lasting excellence.

## CS 110 [AS9100D Clause 8.6] Incomplete Inspection Records: A Repeated NC

During a routine **AS9100 surveillance audit** at AeroTech Industries, the auditor, Bernard Fields, was reviewing the **inspection process** as part of her inspection of production and quality assurance activities. While going through the inspection records for a batch of components, Bernard identified a significant issue: **NC issued: 8.6**.

Upon examining the **inspection record sheets**, Bernard found that although the **inspection results** appeared to be in order with correct measurements taken against the drawing, there was a glaring issue. The **acceptance criteria** were **not indicated** anywhere on the inspection sheets. While the inspection personnel had confirmed that the measurements were correct according to the drawing, the absence of the drawing's **actual dimensions** on the inspection sheets was a clear gap.

*Clause 8.6: The inspection records must include not only the measured values but also the **acceptance criteria** to ensure the product meets the required specifications.*

*Bernard issued the following NC: The inspection record sheets do not indicate the acceptance criteria. While the measurements are correct, the required drawing dimensions are not documented on the inspection records, making it unclear whether the product conforms to the specified requirements.*

### Initial Response and Corrective Actions

The quality team at AeroTech Industries immediately acknowledged the issue. The inspector explained that while the correct measurements were recorded during the inspection, he had simply forgotten to write down the **drawing dimensions** as part of the documentation. He admitted that he had verified the measurements against the drawing but missed the key step of including the acceptance criteria in the record.

### Containment Actions

- Immediate Correction: The inspection team retroactively reviewed all inspections carried out for the batch and updated the records with the correct acceptance criteria—the drawing dimensions.
- Internal Notification: A communication was sent to all relevant inspection staff, reminding them to include all required information, including acceptance criteria, in the records.

### RCA

The root cause was identified as a lack of a standardized checklist or procedure that ensured all critical information, including the **acceptance criteria**, was always included in the inspection records. This oversight was attributed to insufficient **documentation training** for new inspection staff and the absence of a formal **final inspection review** process to catch such errors before the records were finalized.

### Corrective Actions

- **Standardized Inspection Template:** A new inspection **record template** was introduced, which clearly listed fields for both the **measured values** and the **acceptance criteria** (drawing dimensions).
- **Training:** The inspection team underwent training on the new inspection templates and the importance of documenting acceptance criteria, particularly the drawing dimensions, as part of the inspection record.

- **Final Review Process:** A **final inspection review** was instituted where a second staff member would verify that all required information, including acceptance criteria, was present before the inspection records were signed off.

Bernard reviewed these corrective actions and agreed that they were appropriate. The **NC was closed** with the understanding that AeroTech would implement these changes and ensure no recurrence.

### First Follow-Up Audit: A Year Later

Bernard returned for the next **surveillance audit** a year later to verify the effectiveness of the corrective actions. She once again reviewed the **inspection records**, expecting to find evidence that the new processes were working. However, she discovered that despite the previous corrective actions, the **same issue** had not been fully resolved.

The inspection records still indicated only **tick marks** or **accepted signatures**, without clearly documenting the **acceptance criteria**—the required drawing dimensions. Bernard was disappointed to find that the corrective actions from the previous audit had not been fully effective in addressing the root cause of the problem.

### New Major NC Issued

**Major Nonconformance:** Despite the corrective actions taken last year, the inspection records still fail to document the acceptance criteria, including the drawing dimensions, as required by the AS9100 standard. This repeated failure to maintain proper records represents a **major** noncompliance with **clause 8.6**.

Bernard rejected the **corrective action report** from the previous year and issued a **major NC** to emphasize the severity of the issue.

### Second Round of Corrective Actions

This time, AeroTech Industries took a more aggressive approach to address the issue. The corrective actions were revised as follows:

1. Revision of Inspection Process:
   - The inspection procedure was restructured to mandate that all inspection records include not only the measurements but a clear reference to the drawing dimensions. This change ensured that inspectors could not overlook this essential step in the process.
   - A checklist was introduced to be completed at the end of each inspection, confirming that all necessary fields were filled out, including acceptance criteria.
2. Management Oversight:
   - The company appointed a senior quality manager to perform periodic checks on the inspection records, ensuring compliance with the updated process.
   - A quarterly internal audit was implemented to ensure that every inspection record met the full documentation requirements, including clear evidence of acceptance criteria.
3. Additional Training and Refresher Courses:
   - The company held refresher training courses for all inspection staff to reiterate the importance of documenting acceptance criteria and the potential impact on product conformity and certification.

### Final Verification and Certification Recommendation

Bernard returned for the follow-up audit **another year later**. This time, the inspection records were thoroughly reviewed and found to be in full compliance with AS9100 requirements. The records now consistently included the necessary **acceptance criteria**, and there was clear evidence of management oversight and regular internal audits.

Bernard confirmed that the **corrective actions had been fully implemented** and were effective in resolving the issue. She **closed the major NC** and recommended that AeroTech Industries' **AS9100 certification** be continued, with confidence in their improved processes.

### Conclusion

After three years, AeroTech Industries finally addressed the root cause of its documentation issues. Through persistent follow-up actions, RCA, and corrective actions, it succeeded in implementing a robust process for **documenting acceptance criteria** for inspections. Bernard's careful oversight ensured that the company was fully compliant with **AS9100** by the end of the final verification.

Chapter 08

# Nonconformities in the Control of Nonconforming Outputs/ Customer Complaints Handling Process

## CS 111 [AS9100D Clause 8.7.1] Aircraft Emergency: Throttle Failure Disrupts Fuel Flow

Let me give you a real story before we discuss the next NC.

Teleflex cable was used for aeroengine throttle control on aircraft, the pilot regulating fuel supply to the engine and speed of revolutions per minute (rpm) of the engine. It was an imported item that required a lead time of one month to get it from the supplier. Overhaul of aircraft is often time-bound. The delivery date for the aircraft was committed by the MRO agency.

While finalizing the assembly of the military trainer aircraft and integrating the engine, the manager in the MRO agency on the shop floor noticed that the Teleflex cable was to be replaced due to damage beyond repairs.

He rejected the cable on the aircraft as it was not reusable and raised a request to purchase the spare from the original supplier with an allowed maximum lead time of one week, costing about $60. The supplier declined the order on the same day "We regret to inform you that we cannot fulfil your order due to the extremely short lead time requested. We can supply it only after two months."

The manager was in a fix. Even if he ordered the spare supply, he could not have got one for the aircraft on the overhaul line. He was an experienced person and remembered removing a Teleflex cable from another aircraft for some reason, which was thrown away in the attic space of the hangar a few years back. He went in search of it and was happy when he got it. He handed over the same to the shop to use it after cleaning. The aircraft was delivered to the training squadron after overhaul.

In one of the training sorties, a trainee pilot took off and did all the training exercises confidently and at full throttle. At the end of the sortie, he was to land back, so he aligned the aircraft to the runway for straight and level flight and tried reducing the speed by moving back the throttle. He was surprised. The throttle was not coming back, and the forward speed of the aircraft and rpm were not getting reduced.

Suddenly an emergency. The pilot went around the airfield talked to the air traffic control (ATC) and explained his predicament. After consultations with the technical team, ATC advised the pilot to switch off/cut off the fuel supply to the engine. The trainee pilot aligned his aircraft to the runway, cut off the fuel supply to the engine, and had a forced landing with great difficulty. Minor damages have happened to the aircraft structure.

A court of inquiry investigated the incident and found that the tube carrying the Teleflex cable got into a geometric lock making it immobile. (The cable passes through a smaller tube which is telescopically retractable into a larger tube. There is adequate overlap ensured between the ends of the smaller tube and the larger tube during manufacture so that under normal operations, the smaller tube does not come out of the larger tube. A geometric lock happens when the smaller tube comes out of the larger tube, making it irreversible and immobile.) The Teleflex cable assembly we know was the one earlier rejected due to inadequate overlap/dimensional deviations on the tube ends. However, the rejected cable assembly had no such marking nor it was physically rendered "unusable" before throwing it in the attic space of the hangar.

*The requirement of AS9100D (clause 8.7.1) reads "Product dispositioned for scrap shall be conspicuously and permanently marked, or positively controlled, until physically rendered unusable."*

The responsible persons were excused because the MRO was not practicing the QMS at the time of the overhaul of the aircraft.

## CS 112 [AS9100D Clause 8.7.1] Scrap No More: AeroDef's Blueprint for Control

AeroDef Solutions is an MRO company specializing in military aircraft support in India. The company is headquartered in Nagpur, the hub of India's aerospace and defense industries, and provides comprehensive services such as **aircraft maintenance,** scheduled and unscheduled maintenance for a range of fighter jets, transport aircraft, and helicopters. Their tagline could be: *Strengthening Wings, Securing Skies.*

In the initial audit of the MRO organization for certification to AS9100/9110, the auditor issued an NC as follows: "Product dispositioned for scrap shall be conspicuously and permanently marked, or positively controlled, until physically rendered unusable. This has not happened in the case of a few items. Only a register is maintained for scrap."

Here's how AeroDef Solutions could address the NC and implement follow-up actions to comply with the requirements of AS9100/9110:

### Follow-Up Actions for the NC

1. Immediate Containment Action:
    - Review all scrap items currently in the facility to identify any that have not been permanently marked or positively controlled.
    - Isolate and secure these items in a designated area with restricted access.
    - Update the scrap register to include detailed tracking of all identified items.
2. RCA:
    - Conduct an RCA using methods like 5 Whys or fishbone diagram to determine why the required marking or control was not implemented.
    - Investigate whether the issue stems from a lack of training, inadequate procedures, or oversight during scrap handling.
3. Corrective Action Plan:
    - Develop and document a corrective action plan to address the root cause.
    - Revise scrap handling procedures to include:
        - Clear instructions for marking items as scrap permanently (e.g., using stamps, engravings, or paint).
        - Guidelines for rendering scrap items physically unusable, such as cutting or crushing components.
    - Implement physical controls such as locked scrap bins or designated storage areas for unmarked items.
4. Training and Awareness:
    - Conduct training sessions for employees involved in scrap handling to ensure they understand the revised procedures and AS9100/9110 requirements.
    - Emphasize the importance of compliance with quality management standards in preserving safety and integrity.

5. Verification and Monitoring:
   - Introduce a system to regularly audit scrap handling and marking processes to ensure compliance.
   - Assign a quality assurance team to monitor adherence and address deviations promptly.
6. Update to the QMS:
   - Revise the QMS documentation to reflect changes in the scrap handling process.
   - Include the updated procedure in internal audit checklists to prevent recurrence.
7. Closure of the NC:
   - Present the updated procedures, training records, and evidence of implementation of actions (e.g., photos of marked scrap and inspection logs) to the auditor during the follow-up audit.
   - Demonstrate how the system now prevents scrap items from being unmarked or uncontrolled.

The auditor, during his verification of NC for closure, verified the corrective actions and reserved his right to verify the effectiveness of corrective actions in the surveillance audit, next year.

By addressing the NC thoroughly, AeroDef Solutions ensured compliance with AS9100/9110 standards requirement and strengthened its commitment to quality and safety.

## CS 113 [AS9100D Clause 8.7.1] Marked for Safety: TechWings' Scrap to Success Story

At a leading certified organization in Malaysia, M/s TechWings Malaysia Sdn Bhd—Penang, renowned for manufacturing high-quality aircraft, a surveillance audit was recently conducted to evaluate adherence to stringent aerospace standards. During the audit, the auditor discovered several items lying unmarked and unsorted in the rejection bay.

*After thorough verification, the auditor issued a* ***major NC*** *under clause* ***8.7.1****: "The organization has not ensured that nonconforming parts are identified and controlled to prevent their unintended use or delivery. Parts disposed of due to deviation are not identified in the rejection bay. Products dispositioned for scrap are not conspicuously and permanently marked, or positively controlled, until physically rendered unusable."*

### The Auditor's Explanation: Consequences of Neglect

The auditor addressed the organization's leadership and quality team, underlining the gravity of the issue by referencing past aviation disasters caused by inadequate handling of nonconforming parts.

1. **The Engine Blade Incident:** In the early 2000s, an airline faced a catastrophic engine failure mid-flight when a nonconforming turbine blade, which was supposed to be scrapped, was mistakenly used during maintenance. The blade fractured under stress, causing a chain reaction that disintegrated the engine and damaged the aircraft's wing. Fortunately, the pilot managed

to land the aircraft safely, but the investigation revealed lapses in identifying and segregating rejected components.

2. **Cabin Decompression Case:** The auditor also shared a chilling account of an incident where a nonconforming fastener, incorrectly classified as fit-for-use, led to cabin decompression at cruising altitude. The structural failure resulted in the loss of life and underscored the devastating consequences of overlooking proper rejection and control measures.

These examples, the auditor emphasized, were stark reminders of why meticulous handling of nonconforming parts was not just procedural but a matter of life and death in the aviation industry.

## Immediate Response: Quality Manager's Initiative

The quality manager at the organization acted promptly to address the NC and prevent future occurrences.

1. **Containment Actions:**
   - The rejection bay was immediately organized, and all nonconforming parts were identified, marked, and segregated.
   - A dedicated team was assigned to ensure no further deviations occurred during the interim period.
2. **RCA:** The investigation identified the following causes:
   - **Human Error:** Personnel handling nonconforming parts lacked adequate training and awareness of the criticality of proper marking and segregation.
   - **Process Gaps:** Procedures for managing nonconforming parts were not comprehensive or rigorously enforced.
3. **Corrective Actions:**
   - **Revised Procedures:** The organization implemented updated SOPs for handling, marking, and segregating nonconforming parts, ensuring clarity and accountability at every step.
   - **Enhanced Training:** All staff involved in quality control and inventory management underwent intensive training focused on the criticality of identifying and managing nonconforming parts.
   - **Permanent Marking:** Scrapped products were conspicuously and permanently marked using automated systems to eliminate any chance of unintended use.
   - **Dedicated Storage:** Nonconforming and scrapped items were moved to a secure area with restricted access, monitored by quality personnel.
4. **Management Oversight:** Periodic audits and real-time monitoring were introduced, with findings reported to senior management.

## Verification on OASIS

The auditor reviewed the corrective actions on the **OASIS** and commended the organization's swift and detailed response. However, the auditor stressed the importance of consistency, stating:

"The effectiveness of these corrective actions will be verified during next year's surveillance audit. Sustained vigilance is essential to ensuring safety and compliance."

## Conclusion: A Culture of Accountability

This incident served as a wake-up call for the organization, underscoring the critical importance of rigorous quality management in the aviation sector. By addressing the NC with urgency and implementing robust corrective actions, the organization reaffirmed its commitment to safety, quality, and compliance. Moving forward, the leadership vowed to foster a culture where every employee recognizes the vital role they play in ensuring the safety and reliability of their products.

## CS 114 [AS9100D Clause 8.7.1] Bad Parts: A Sky-High Tale of Accountability

At a leading aircraft manufacturing organization in Malaysia, an auditor conducted a surveillance audit to assess compliance with aerospace quality standards. During the audit, while reviewing the control of nonconforming products, the auditor discovered a serious oversight.

A nonconforming part had been disposed of by an inspection personnel without seeking the required approval from their manager. Upon further investigation, the auditor found that the organization's procedure for controlling nonconforming products failed to clearly define the **responsibility and authority for the review and disposition of nonconforming outputs**. This gap led to a violation of clause **8.7.1**, resulting in the issuance of a**n NC**:

> *The procedure for control of nonconforming products does not describe the responsibility and authority for the review and disposition of nonconforming outputs. An inspection personnel disposed of the NC part without managerial approval, breaching process control requirements.*

### The Auditor's Perspective: Lessons from the Past

The auditor elaborated on the criticality of the issue, citing his past experiences in the aviation industry where lapses in nonconformance management had dire consequences:

1. **Unapproved Repairs and Catastrophic Failure:** In one case, an unauthorized repair was performed on a nonconforming component of an aircraft's hydraulic system. Without proper review and approval, the component was cleared for installation. During the flight, the system failed, leading to a loss of control and a forced emergency landing. While no lives were lost, the incident underscored the importance of ensuring all nonconforming products are reviewed and approved by qualified personnel.
2. **Supply Chain Mismanagement:** The auditor also recounted an incident involving a subcontractor who bypassed proper review protocols to meet delivery deadlines. A defective batch of fasteners made its way into critical assemblies, causing a structural failure during a routine test flight. Investigations revealed the lack of clear accountability for approving and tracking nonconforming outputs, leading to significant financial and reputational damage.

The auditor emphasized:

> *Clear accountability and authority are essential in handling nonconforming products. The absence of these controls can compromise safety, lead to operational failures, and tarnish organizational credibility.*

## Quality Manager's Response

Acknowledging the gravity of the NC, the quality manager initiated a structured response to address the issue comprehensively:

1. **Containment Actions:**
   - All nonconforming products in the facility were immediately inventoried, and their status was verified to ensure no further unauthorized actions had been taken.
   - A temporary approval workflow was implemented, requiring dual authorization for any disposition of nonconforming parts until a permanent solution was in place.
2. **RCA:** The RCA identified the following key issues:
   - **Procedural Gap:** The existing procedure lacked clear documentation of roles, responsibilities, and authority for the review and disposition of nonconforming outputs.
   - **Human Error:** The inspection personnel, unaware of the need for managerial approval, acted independently.
   - **Inadequate Training:** Staff had not received sufficient training on the importance of adhering to nonconformance protocols.
3. **Corrective Actions:**
   - **Procedure Update:** The control of nonconforming products procedure was revised to clearly define:
     - The roles and responsibilities for nonconforming product review.
     - The approval hierarchy for dispositioning nonconforming products.
   - **Staff Training:** Comprehensive training sessions were conducted for all employees, emphasizing the criticality of the revised procedures and the potential risks of noncompliance.
   - **System Enhancement:** An automated system was implemented to flag any unauthorized nonconforming product actions, requiring managerial approval for final disposition.
   - **Quality Oversight:** A quality oversight committee was established to monitor and audit nonconforming product handling regularly.
4. **Management Communication:** Senior management emphasized their commitment to quality and safety, ensuring all employees understood the importance of adhering to the updated procedures.

## Auditor's Verification

The auditor reviewed the corrective actions on the **OASIS** and noted significant improvements in the organization's approach to managing nonconforming products. He commended the quality manager for the swift and thorough response but cautioned: "The true test of these corrective actions lies in their consistent implementation. I will verify their effectiveness during next year's surveillance audit."

## Conclusion: Embedding a Culture of Quality

This NC served as a valuable lesson for the organization, highlighting the need for robust systems to manage nonconforming products and ensure safety in every aspect of production. The incident catalyzed a cultural shift, driving the organization to prioritize accountability and continual improvement. By embedding these values, the organization reinforced its reputation as a trusted manufacturer in the aerospace industry.

# CS 115 [AS9100D Clause 8.7.1] Strengthening Authorization: Fixing Critical Nonconformance

At an AS9100-certified aircraft manufacturing organization in China, a routine surveillance audit was conducted to ensure compliance with aerospace quality management standards. During the audit, the auditor identified a significant nonconformance related to the disposition of nonconforming products.

While dispositions such as **"use-as-is"** or **"repair"** had been implemented for nonconforming products, the auditor discovered that the manager approving these dispositions lacked the required **authorization or delegation of authority** from the design organization, customer, or regulatory authority. This oversight violated clause **8.7.1** and was flagged as a serious gap in compliance.

The auditor issued the NC with the following observation:

*Dispositions of use-as-is or repair for the acceptance of nonconforming products have been implemented. However, there is no authorization or delegation of authority from the design organization, customer, or regulatory authority available to the manager who gives the disposition.*

## The Auditor's Perspective: Lessons from Industry

To emphasize the gravity of the issue, the auditor shared examples from his extensive experience in aerospace quality assurance:

1. **Unauthorized Repairs Leading to Failure:** In a previous case, a repair was performed on a structural component without proper authorization from the design organization. Though the component passed initial inspections, it failed during operational stress, leading to an emergency landing. Investigations revealed that the lack of authorized approval for the repair was a key contributor to the failure.
2. **Compromised Safety in Use-As-Is Decisions:** The auditor also highlighted an incident where a "use-as-is" disposition for a nonconforming fuel system component was made without proper authority. During flight, the component malfunctioned, causing operational disruptions. The root cause is traced back to the absence of design organization approval for the disposition.

The auditor stressed the importance of adhering to authorization protocols:

*In the aerospace industry, every decision regarding nonconforming products must be backed by appropriate authority to ensure safety, compliance, and reliability. Deviations can have catastrophic consequences.*

## The Quality Manager's Response

Recognizing the criticality of the NC, the quality manager promptly initiated a structured plan to address the issue and establish robust controls:

1. **Containment Actions:**
   - All current dispositions of nonconforming products were reviewed to ensure compliance with authorization requirements.

- Any nonconforming products with unapproved "use-as-is" or "repair" dispositions were flagged and quarantined pending proper approval.

2. **RCA:** The RCA identified the following gaps:
    - **Lack of Authorization Framework:** No formal system was in place to obtain or document authorization from the design organization, customer, or regulatory authority.
    - **Inadequate Training:** Managers lacked awareness of the requirement for external authorization for specific dispositions.
    - **Process Oversight:** Internal procedures did not mandate the verification of authorization before implementing dispositions.
3. **Corrective Actions:**
    - **Procedure Revision:** The organization's procedures for handling nonconforming products were updated to include mandatory steps for obtaining and documenting authorization or delegation of authority from the design organization, customer, or regulatory authority before approving "use-as-is" or "repair" dispositions.
    - **Authorization Records:** A centralized system was implemented to maintain records of all authorizations and delegations, ensuring traceability and compliance.
    - **Training Programs:** Comprehensive training sessions were conducted for managers and relevant staff to reinforce the importance of obtaining proper authorization and understanding the revised procedures.
    - **Approval Workflow:** An automated workflow was introduced, requiring documented authorization to be attached before finalizing any disposition for nonconforming products.
4. **Monitoring and Oversight:** Periodic audits were scheduled to verify compliance with the updated procedures and ensure no dispositions were made without proper authorization.

### Auditor's Follow-Up

The auditor reviewed the corrective actions on the OASIS and noted significant improvements. He acknowledged the quality manager's comprehensive response, stating: "The corrective actions address the root cause effectively. However, their sustained implementation and effectiveness will be verified during next year's surveillance audit."

### Conclusion: Elevating Quality and Safety Standards

This NC served as a critical reminder of the importance of strict adherence to authorization protocols in the aerospace industry. By taking swift and decisive action, the organization not only resolved the immediate issue but also reinforced its commitment to quality, safety, and compliance. Moving forward, the organization vowed to embed these practices into its culture, ensuring the safety and reliability of its products.

## CS 116 [AS9100D Clause 8.7.1] Soaring To Compliance: A Race to AS9100D Certification

Lee, the dedicated Production Manager of a mid-sized private industry in Singapore, had a daunting challenge ahead. The company had recently expanded into manufacturing subassemblies for aircraft, a promising but highly regulated market. Corporate leadership set an ambitious goal: achieve AS9100D certification within just three months.

The CEO, a seasoned automotive industry veteran well versed in ISO 9001, led the effort. However, his knowledge of ISO 9001 did not fully prepare him for the nuanced and stringent requirements of AS9100D, an aerospace-specific standard with zero tolerance for process lapses. In preliminary meetings, the team drew up a certification plan, unaware that critical requirements—particularly around nonconforming products—were overlooked.

Despite the gaps, the team worked tirelessly, believing they were ready when the stage 2 audit began. The lead auditor, Dr. Priyadarshanan, a veteran from India with deep aerospace expertise, meticulously reviewed processes. His inspection zeroed in on the control of nonconforming outputs, where he discovered a major nonconformance under clause 8.7.1.

The finding was serious: the design liaison engineer, without delegated authority from the D&D agency, had approved deviations, authorized "use-as-is" dispositions, and cleared repairs—all without proper design or customer approval. Clause 8.7.1 explicitly mandated that such dispositions required authorization from an official design representative or the customer if deviations impacted contract requirements.

"Delivery of final products as per ISO 9001 practice is not acceptable for aerospace equipment," Dr. Priyadarshanan explained, emphasizing the life-critical nature of aerospace components.

The discovery was a severe setback, but Lee was determined to turn things around. Gathering his team, he collaborated closely with the design department to implement swift corrective actions. The design liaison engineer's authority was formally redefined, and a proper delegation system was established. Only personnel with explicit design approval could make decisions on nonconforming products.

They also introduced new training sessions to deepen employee understanding of AS9100D, along with regular internal audits to catch potential issues early. The revised system was airtight.

Upon re-audit, the improvements were evident. Dr. Priyadarshanan validated the corrections, and the initial AS9100D certificate was granted. The company's perseverance had paid off.

The benefits were transformative:

- Enhanced Credibility: The certification opened doors to high-value aerospace contracts.
- Robust Processes: The strengthened QMS minimized production errors.
- Customer Trust: Delivery of defect-free, fully compliant products boosted client confidence.
- Continual Improvement Culture: The challenge instilled a proactive mindset, driving ongoing process optimization.

Lee stood on the factory floor, watching as a completed subassembly was carefully packaged for shipment. They had navigated turbulence, but now the company was ready to soar—stronger, more disciplined, and certified to fly.

## CS 117 [AS9100D Clause 8.7.1] Fixing Oversights in Corrective Verifications: A Critical NC

At an AS9100-certified aircraft manufacturing organization in China, a surveillance audit was conducted to evaluate compliance with aerospace industry quality management standards. During the audit, the auditor identified a critical nonconformance related to the verification of corrections for nonconforming outputs.

The auditor observed that **"conformity to the requirements shall be verified when nonconforming outputs are corrected"** (as per clause **8.7.1**) was not evident in certain snag sheets from the final assembly of aircraft and aeroengines. Specifically, while issues noted during the final assembly had been corrected, there was no documented evidence of verification to confirm that the corrections met the required specifications.

This oversight was flagged as a significant nonconformance:

> *Conformity to the requirements shall be verified when nonconforming outputs are corrected. This verification is not evident in some snag sheets for the final assembly of aircraft/aeroengines.*

### The Auditor's Perspective: Lessons from Past Incidents

To emphasize the importance of verifying corrections, the auditor shared his experiences from past audits:

1. **Unverified Corrections Leading to Flight Delays:** In one instance, a nonconforming wiring harness in an aircraft's avionics system was repaired without proper verification. During a pre-delivery flight test, the system malfunctioned, causing delays in delivery and additional costs for rework. The root cause was traced back to inadequate verification of the corrected output.
2. **Catastrophic Failure from Overlooked Verifications:** The auditor also recounted a case where a nonconforming landing gear component was repaired but not properly verified before installation. During landing, the component failed, leading to significant damage to the aircraft. This incident underscored the importance of rigorous verification processes to ensure safety and reliability.

The auditor stressed: "Verification of corrections is a cornerstone of quality management. In the aerospace industry, overlooking this step can lead to operational failures, jeopardize safety, and erode customer trust."

### The Quality Manager's Response

The quality manager took immediate action to address the NC and reinforce the organization's commitment to quality and safety:

1. **Containment Actions:**
   - A review of all snag sheets from recent final assemblies was conducted to identify gaps in verification records.
   - Nonconforming outputs with unverified corrections were flagged for immediate re-inspection and verification.

2. **RCA:** The RCA revealed the following issues:
    - **Documentation Gaps:** The process for recording verification of corrections was inconsistently applied.
    - **Human Error:** Some team members lacked awareness of the requirement to document verification after corrections.
    - **Inadequate Oversight:** Supervisors failed to ensure that verification steps were completed and recorded.
3. **Corrective Actions:**
    - Process Update:
        - The procedure for handling nonconforming outputs was revised to include mandatory steps for verification and documentation.
        - A checklist was introduced to ensure verification was completed and signed off for each corrected output.
    - Training and Awareness:
        - Comprehensive training sessions were conducted to educate employees on the importance of verification and the updated procedures.
        - Supervisors received additional training on monitoring and enforcing compliance with verification requirements.
    - System Enhancements:
        - An automated system was implemented to track nonconforming outputs, corrections, and verification records. This system flagged incomplete steps to prevent unverified corrections from progressing.
4. **Management Oversight:**
    - Weekly reviews of snag sheets and verification records were introduced to ensure compliance.
    - Internal audits were scheduled to monitor the effectiveness of the corrective actions.

### Auditor's Follow-Up

The auditor reviewed the corrective actions on the **OASIS** and commended the organization for its comprehensive approach. However, he cautioned: "While the corrective actions address the immediate issue, their sustained effectiveness will be evaluated during next year's surveillance audit. Ensuring consistent implementation is key to maintaining compliance."

### Conclusion: A Culture of Accountability

This NC highlighted the critical role of verification in the quality management process. By addressing the gaps, the organization reinforced its dedication to maintaining the highest standards of safety and reliability. The incident served as a reminder to all employees of the importance of thorough documentation and verification, laying the foundation for a stronger culture of accountability and continual improvement.

## CS 118 [AS9100D Clause 8.7.2] Fixing NCR Retention and Concession Oversights: A Critical NC

At an AS9100-certified aircraft manufacturing organization in China, an auditor conducted a surveillance audit to assess compliance with aerospace quality standards. During the review, a significant NC was identified related to the retention of NCRs and the management of concessions.

The auditor observed that the organization was not retaining NCRs for parts where concessions were granted. Furthermore, the concession format did not require recording the root cause or corrective actions, leading to a repetitive cycle of issuing concessions without addressing underlying issues. This practice violated the requirements of clause **8.7.2**, which states:

*NCR for a part shall have details of the NC, shall describe any concessions obtained, and shall identify the authority deciding the action in respect of the NC.*

*The auditor issued the NC with the following words: "No NCR is retained where a concession is obtained. Corrective actions are not being implemented, and concessions are repeatedly raised. The current concession format does not require documenting root causes or corrective actions" (clause 8.7.2).*

### The Auditor's Perspective: Risks of Inadequate Concession Management

The auditor explained the critical risks associated with this practice, drawing on his extensive experience:

1. **Repeat NCs:** In a previous audit at another organization, a recurring concession for a structural component defect was granted without addressing the root cause. This eventually led to the defect appearing in multiple batches, resulting in costly recalls and rework.
2. **Compromised Quality and Safety:** The auditor recounted an incident where an improperly documented concession allowed a nonconforming fastener to be used in an aircraft assembly. Without corrective actions, the fastener failed during operation, posing significant safety risks.
3. **Regulatory Noncompliance:** The absence of documented NCRs and corrective actions undermines compliance with regulatory and customer requirements, risking penalties, reputational damage, and loss of certification.

The auditor emphasized: "Retaining NCRs, documenting concessions, and addressing root causes are not just procedural requirements—they are essential to ensuring quality, safety, and compliance in aerospace manufacturing."

## The Quality Manager's Response

The quality manager took immediate action to address the NC and prevent recurrence:

**1. Containment Actions:**

- A review of all concessions issued in the past year was conducted to identify and document details of the related NCs.
- Any active concessions were flagged for additional scrutiny to ensure they did not pose immediate safety or quality risks.

**2. RCA:** The RCA revealed the following gaps:

- **Process Gaps:** No clear linkage between NCRs and concessions in the existing system, leading to missing documentation.
- **Lack of Oversight:** Approvals for concessions were granted without evaluating the need for corrective actions.
- **Cultural Mindset:** A tendency to repeatedly issue concessions as a workaround, rather than addressing the underlying problem.

**3. Corrective Actions:**

- Process Revision:
  - The NCR process was revised to mandate that every concession must reference a corresponding NCR.
  - The concession format was updated to include fields for recording the root cause, corrective actions, and approval authority.
- Training Programs: Training sessions were conducted for all relevant employees to reinforce the importance of documenting NCRs, addressing root causes, and implementing corrective actions.
- Approval Workflow: A stricter approval workflow was implemented, requiring higher-level management or design organization approval for repeated concessions.
- Monitoring and Reporting: A monthly report was introduced to track and analyze trends in NCRs and concessions, ensuring corrective actions are effectively reducing NCs.

**4. Management Oversight:**

- Regular management reviews were initiated to evaluate the effectiveness of the revised NCR and concession processes.
- Internal audits were scheduled to verify compliance and identify any further improvements.

## Auditor's Follow-Up

The auditor reviewed the corrective actions on the OASIS and commended the organization for its comprehensive approach. However, he noted: "The corrective actions address the root cause effectively. I will verify their sustained implementation and effectiveness during the next surveillance audit."

## Conclusion: Building a Culture of Continual Improvement

This NC highlighted the importance of retaining NCRs and using concessions judiciously to drive corrective actions and prevent recurring issues. By taking decisive steps, the organization not only resolved the immediate nonconformance but also strengthened its commitment to quality and safety. Moving forward, the organization aims to embed these improvements into its culture, ensuring long-term compliance and excellence in aerospace manufacturing.

Chapter 09

# Nonconformities in Performance Evaluation

## CS 119 [AS9100D Clause 9.1.1] The Audit Alarm: A Path to Improvement

It was a sunny Monday morning when the certification audit for AS9100D commenced at the software organization—a reputed firm providing design services for a prime aircraft manufacturer. The team was confident, as their processes had always aligned closely with the aerospace industry standards. However, midway through the audit, the air in the conference room turned tense.

The auditor, a seasoned professional with decades of experience, had uncovered a NC in management review processes. With a calm but firm demeanor, he stated, **"There is no evidence to show that the performance data of the organization is being monitored, analyzed, and evaluated for improvement. There are no records retained."**

The announcement was met with a mix of surprise and concern. Management Representative Arjun Kumar tried to explain. "We've been focusing heavily on meeting design deadlines and delivering quality outputs to our clients. Unfortunately, we do not have adequately trained personnel to collect and analyze performance data," he admitted.

The auditor raised an eyebrow and added pointedly, "The current practices of your QMS do not match your reputation in the field. You can definitely improve."

He emphasized the necessity of data analysis, elaborating on its role in identifying trends, minimizing inefficiencies, and driving continual improvement. "This is not just about compliance—it is about staying competitive in a demanding industry. Monitoring and analyzing performance data can uncover insights that strengthen your position."

Before concluding the audit, the auditor issued clear instructions. "Containment actions are required immediately. You must also conduct an RCA to identify the underlying issues. Implement corrective actions, and ensure records are maintained for future reference."

### The Path to Recovery

The days following the audit were intense. The software organization formed a dedicated team to address the NC. They began by identifying KPIs relevant to their design services. These included project turnaround times, defect rates, and client satisfaction scores.

They also conducted an RCA, uncovering gaps in training and a lack of a structured framework for data collection and evaluation. In response, the firm implemented the following corrective actions:

1. **Training and Development:** Employees across departments underwent training in performance data analysis and record management.
2. **System Integration:** The firm integrated performance data collection modules into its project management software, ensuring automated tracking and analysis.
3. **Regular Reviews:** A schedule for monthly performance reviews was established, where trends and improvement opportunities were discussed with leadership.

To ensure the effectiveness of these actions, all changes were meticulously documented and uploaded to the OASIS, as required.

### One Month Later

A month after the initial audit, the auditor returned for a follow-up. Armed with a checklist, he meticulously reviewed the organization's corrective actions. The uploaded evidence from OASIS demonstrated a clear commitment to improvement. The auditor noted the implementation of the new system, training records, and meeting minutes from recent performance reviews.

"Well done," the auditor remarked after his verification. "You've taken commendable steps to address the NC. However, I will verify the effectiveness of these corrective actions during the upcoming surveillance audit. Ensure that the momentum is sustained."

The team breathed a collective sigh of relief, but they knew the work was not over. The NC had been a wake-up call, reminding them that even a well-established organization must continually strive for excellence.

### A Lesson Learned

The experience served as a turning point for the organization. They began leveraging data analytics not just for compliance, but as a tool for driving innovation and operational efficiency. Over time, the enhanced processes reflected higher client satisfaction and a stronger reputation in the aerospace sector.

The auditor's final words resonated deeply: "Improvement is a journey, not a destination. Make data your ally."

## CS 120 [AS9100D Clause 9.1.2] The Missing Link in Customer Satisfaction

The certification audit for AS9100D was in full swing at a well-regarded software organization specializing in design services for a prime aircraft manufacturer. As the auditor progressed through the processes, he paused during the review of management review processes. After meticulous scrutiny, he pointed out a significant NC.

> *"Clause 9.1.2: Customer satisfaction determination is not effective," the auditor stated firmly. "The survey method used for obtaining, monitoring, and reviewing this information often does not reflect customers' perceptions of the degree to which their needs and expectations have been fulfilled. Survey sheets from customers are filed, and no further actions are evident."*

The statement left the room in silence, broken only by the nervous shuffle of papers. The Management Representative, Priya Mehta, stepped forward to explain. "We frequently have customer visits, and they usually leave satisfied. Additionally, we address complaints promptly. Customers also provide us with scorecards. Is not that sufficient to demonstrate their satisfaction?"

The auditor gave a measured nod before responding, "While customer visits and scorecards are valuable, they do not provide a comprehensive view of how well their needs and expectations are being met. Effective customer satisfaction determination goes beyond filing survey sheets. It involves analyzing the data, identifying trends, and taking proactive actions to enhance customer experiences."

He continued, emphasizing the importance of a structured approach to customer satisfaction determination. "A robust method of assessing customer perceptions is essential—not only for compliance but also to strengthen your relationship with clients and your standing in the industry."

Before concluding the discussion, the auditor requested the Management Representative to initiate containment actions, conduct an RCA, and implement corrective measures. The room buzzed with activity as the team began to address the issue head-on.

### The Road to Rectification

In the days following the audit, the organization took decisive steps to resolve the NC. The quality management team, with support from leadership, worked tirelessly to develop and implement an improved approach to customer satisfaction determination. The following corrective actions were put in place:

1. **Enhanced Survey Mechanism:** The team redesigned customer satisfaction surveys to capture detailed feedback on various performance metrics, including timeliness, quality, communication, and support.
2. **Data Analysis Framework:** A dedicated team was tasked with analyzing survey responses, identifying trends, and highlighting areas for improvement.
3. **Actionable Insights:** Survey results were no longer filed away. Instead, they were reviewed in monthly management meetings, and actionable insights were documented with follow-up plans.
4. **Customer Feedback Loop:** A formal process was established to communicate back to customers about actions taken based on their feedback, reinforcing the organization's commitment to continual improvement.

All changes were meticulously documented and uploaded to the OASIS, as required.

### A Month Later

The auditor returned a month later for a follow-up audit. With a checklist in hand, he reviewed the implemented corrective actions. The redesigned survey forms, detailed analysis reports, and documented meeting minutes were presented as evidence. The auditor scrutinized the records and acknowledged the efforts.

"You've taken meaningful steps to address the NC," he noted. "The documentation and new processes reflect a genuine commitment to understanding and fulfilling customer needs. However, the true test will be during the surveillance audit, where we'll assess the effectiveness of these actions over time."

The team felt a sense of accomplishment but understood the importance of sustaining their efforts.

### A New Perspective

The experience served as a pivotal moment for the organization. By embracing a proactive and structured approach to customer satisfaction determination, they not only resolved the NC but also strengthened their relationships with clients. Customer feedback became an invaluable tool for innovation and growth.

As the auditor prepared to leave, he offered a final piece of advice: "Customer satisfaction is not just a metric—it reflects your organization's commitment to excellence. Continue to listen, learn, and act, and you'll set a benchmark in the industry."

The lesson was clear: compliance was not merely a checkbox but a pathway to building trust and achieving long-term success.

## CS 121 [AS9100D Clause 9.1.3] Gap in the Analysis: Turning Data into Action

The certification audit for AS9100D was underway at a software organization renowned for its design services for a prime aircraft manufacturer. The audit was proceeding smoothly until the auditor began reviewing management review processes. As he delved deeper into the evidence, his expression grew serious.

*"Clause 9.1.3: Analysis of results of measurement used to evaluate the effective implementation of planning is not evidenced," the auditor declared.*

The room fell silent. The Management Representative, Rajesh Kapoor, stepped forward, attempting to explain. "We do measure our data—delivery timelines, defect rates, resource utilization. But... we haven't consolidated or formally analyzed it to evaluate the effectiveness of our planning. We're focused on operational tasks, and honestly, we haven't had the bandwidth."

The auditor listened patiently and then responded firmly. "Measurement without analysis is like having a compass but refusing to look at it. The purpose of measurement is to provide insight into your performance and guide improvements. Without evaluating the effectiveness of your planning, you risk repeating inefficiencies or missing opportunities for optimization."

He elaborated on the necessity of analyzing measured data to ensure that planning processes were delivering the intended results. "This is not just about compliance; it is about improving operational efficiency, resource allocation, and customer satisfaction. You must take this seriously."

The auditor instructed the Management Representative to initiate containment actions, conduct an RCA, and develop a plan to implement corrective actions.

## Addressing the NC

Determined to address the NC, the organization formed a task force comprising members from project management, quality assurance, and operations. The team began by assessing the gaps in their current processes.

The RCA revealed:

- **Fragmented Data Management:** While data were being collected, they were not systematically consolidated or analyzed.
- **Lack of Ownership:** No single department or individual was responsible for performance analysis.
- **Skill Gaps:** Team members lacked expertise in data analysis and its application to planning evaluation.

With these insights, the organization implemented the following corrective actions:

1. **Centralized Data Repository:** A centralized system was developed to collect, store, and manage performance data, ensuring easy access for analysis.
2. **Performance Analysis Framework:** A structured approach was established to analyze data, focusing on delivery schedules, defect rates, resource efficiency, and customer satisfaction metrics.
3. **Ownership and Accountability:** A dedicated team was assigned responsibility for conducting regular performance analyses and reporting findings during management reviews.
4. **Training and Development:** Key personnel underwent training on data analysis techniques and tools to enhance their capability to evaluate planning effectiveness.

The team documented every step of their corrective actions and uploaded the evidence to the OASIS, ensuring transparency and readiness for the follow-up audit.

## The Follow-Up Audit

A month later, the auditor returned to verify the corrective actions. During the review, the organization presented:

- Reports demonstrating comprehensive analysis of measured data.
- Meeting minutes from management reviews discussing the effectiveness of planning.
- Action plans developed based on analytical insights, including adjustments to resource allocation and timelines.

The auditor reviewed the evidence carefully and nodded in approval. "You've made significant progress. These actions demonstrate your commitment to not only addressing the NC but also improving your operational processes."

However, he added a word of caution. "The true test will be during the surveillance audit, where we'll evaluate the sustained effectiveness of these measures. Ensure that this becomes a consistent practice."

### Lessons and Transformation

The NC had been a wake-up call for the organization, highlighting the importance of data-driven decision-making. By integrating performance analysis into its routine processes, the company not only addressed the audit findings but also unlocked new opportunities for optimization and growth.

As the audit concluded, the auditor left the organization with a parting thought: "Planning without analysis is like flying without navigation. Use your data to guide your path and achieve excellence."

The experience reinforced the organization's resolve to build a culture of continual improvement, where measurement, analysis, and action worked in harmony to drive success.

## CS 122 [AS9100D Clause 9.1.3] Risk Without Reflection: A Lesson in Mitigation

The AS9100D certification audit at a sub-assembly manufacturing organization supplying to a prime aircraft manufacturer was progressing uneventfully until the auditor began scrutinizing management review processes. His sharp eye for detail soon uncovered a significant gap.

> *"Clause 9.1.3: Analysis of results of measurement used to evaluate the effectiveness of actions taken to address risks and opportunities is not evidenced," he announced firmly.*

Management Representative Suresh Nair shifted uneasily in his chair before offering an explanation. "We have risk assessment matrices and mitigation plans. We address issues as they arise, and we haven't had any major failures. Our focus has been more on getting the production out on time."

The auditor raised an eyebrow and replied, "Addressing risks without measuring and analyzing the effectiveness of your actions is like flying blind. You may be resolving individual issues, but without evaluation, how do you know whether your strategies are truly effective or whether you're improving overall?"

He elaborated on the importance of evaluating actions taken to address risks and opportunities, linking it to improved performance, reduced operational disruptions, and better decision-making. "Data analysis helps identify whether the risks were mitigated effectively and whether new opportunities have been capitalized upon. It is an essential part of continual improvement."

Before concluding the audit, the auditor directed the Management Representative to initiate immediate containment, conduct an RCA, and implement corrective actions.

### Addressing the NC

Determined to address the NC, the organization formed a cross-functional team to evaluate its risk management processes. The RCA revealed:

- **Reactive Approach:** Risks were identified, but there was no follow-up to measure the success of mitigation actions.
- **Lack of Metrics:** No quantitative or qualitative metrics were defined to evaluate the effectiveness of risk-related actions.
- **Siloed Processes:** Different teams were addressing risks independently without a unified framework for evaluation.

Based on these findings, the organization implemented the following corrective actions:

1. **Risk Effectiveness Metrics:** Developed a set of metrics to evaluate risk mitigation actions, including frequency of recurrence, cost impact, and downtime reduction.
2. **Unified Risk Dashboard:** Introduced a centralized dashboard for tracking risks, mitigation actions, and their outcomes across all departments.
3. **Performance Reviews:** Incorporated risk and opportunity evaluations as a regular agenda item in monthly management reviews.
4. **Training on Risk Analysis:** Conducted training sessions to educate employees on evaluating the effectiveness of risk mitigation and opportunity actions.

All corrective actions were documented and uploaded to the OASIS for review.

### Follow-Up Audit

A month later, the auditor returned to verify the corrective actions. The organization presented:

- Evidence of risk metrics tracking and analysis.
- Updated management review minutes demonstrating the discussion of risks, opportunities, and mitigation effectiveness.
- A case study showing how a previously recurring risk was effectively mitigated using the new process.

The auditor reviewed the documentation and expressed his approval. "You've established a structured process for evaluating risks and opportunities. These corrective actions not only resolve the NC but also strengthen your operational resilience."

However, he reminded them, "The true test lies in consistency. During the surveillance audit, I will assess how well these practices are sustained and their impact on your overall performance."

### Moving Forward with Confidence

The experience served as a pivotal moment for the organization. By integrating risk evaluation into their regular processes, they were better equipped to identify trends, prevent issues, and seize opportunities.

As the auditor departed, he left them with an insightful thought: "In aviation, risks can't be eliminated, but they can be managed. The key is to constantly evaluate and improve your mitigation strategies."

With renewed focus, the organization committed to making risk evaluation and opportunity management an integral part of its culture, ensuring a stronger future in the aerospace industry.

## CS 123 [AS9100D Clause 9.1.3] A Quality Wake-Up Call

During the AS9100D certification audit, a sub-assembly manufacturing organization supplying to a prime aircraft manufacturer encountered a challenging moment. The auditor, known for his sharp attention to detail, was deep into reviewing management review processes when he paused and made a decisive statement.

*Clause 9.1.3 C: Analysis of results of measurement used to evaluate the need for improvements to the QMS is not evidenced.*

The declaration left the room in a palpable silence. The Management Representative, Kavita Sharma, attempted to justify the situation. "We've been focused on meeting immediate delivery timelines and customer requirements. We do track performance, but analyzing data for improvements to the QMS hasn't been our top priority."

The auditor listened patiently before responding with measured emphasis. "Meeting immediate demands is important, but failing to analyze performance data for systemic improvements puts your long-term efficiency and customer satisfaction at risk. Improvements to the QMS aren't just about addressing immediate issues—they're about building a resilient and efficient system for the future."

He elaborated on the necessity of evaluating the effectiveness of actions and identifying areas where the QMS could be enhanced. "Analysis helps you stay competitive, reduce risks, and seize OFI. It is not optional—it is fundamental to your success. The effectiveness of your QMS is measured by the number of NCs you get during internal audits. Do a trend analysis to see the results whether you are improving or not."

Before moving forward, the auditor requested Sharma to initiate containment, perform an RCA, and implement corrective actions.

### The Journey to Improvement

Recognizing the urgency of the NC, the organization assembled a cross-functional team to address the gap. A comprehensive RCA was conducted, uncovering key issues:

- **Reactive Focus:** Efforts were concentrated on addressing customer complaints and production delays, rather than proactive QMS improvements.
- **Lack of Analytical Tools:** The absence of structured tools and methods to analyze QMS performance data hindered systematic evaluations.
- **No Assigned Responsibility:** There was no dedicated individual or team accountable for identifying QMS improvement needs.

Armed with these insights, the team implemented the following corrective actions:

1. **Systematic Data Analysis:** Introduced a structured framework for analyzing data from audits, customer feedback, production metrics, and corrective actions to identify trends and areas for QMS improvement.
2. **Defined Metrics for Improvement:** Established KPIs for the QMS, including OTD, defect reduction, and customer satisfaction scores.

3. **Regular Review Cycles:** Incorporated QMS improvement evaluations as a standing agenda item in management reviews, ensuring a consistent focus on systemic enhancements.
4. **Ownership and Accountability:** Assigned a dedicated QMS improvement officer to oversee the process of identifying and implementing improvements.
5. **Training and Awareness:** Conducted training sessions for employees to understand the importance of QMS improvements and their role in supporting this objective.

All actions and associated evidence were meticulously documented and uploaded to the OASIS for review.

## Follow-Up Audit

One month later, the auditor returned to assess the organization's progress. During the follow-up audit, the team presented:

- Reports of analyzed performance data, showing identified trends and proposed improvement actions.
- Management review minutes, documenting discussions and decisions related to QMS enhancements.
- Evidence of implemented changes, such as updated processes and additional training for employees.

The auditor reviewed the documentation and observed the updates with satisfaction. "You've demonstrated a thorough approach to addressing this NC. These actions reflect a commitment to continual improvement," he remarked.

However, he added a word of caution: "The real measure of success lies in sustainability. I'll assess the effectiveness of these corrective actions during the surveillance audit. Keep this momentum going."

## A Culture of Continual Improvement

The NC served as a catalyst for transformation within the organization. By embedding systematic analysis and improvement practices into their QMS, they not only resolved the NC but also established a culture of proactive improvement.

As the auditor left, he shared a final thought: "Improvement is not just about fixing what's broken—it is about making your processes better every day. That's how you build a legacy of excellence."

The organization embraced this philosophy, ensuring that its QMS became a cornerstone of their success in the competitive aerospace industry.

## CS 124 [AS9100D Clause 9.2.1] Internal Audit Review: A Production Manager's Journey to AS9100D Compliance

The clock was ticking. Three months—that was the deadline given by corporate leadership to achieve AS9100D certification for the newly established aerospace subassembly division. For Rajesh Nair, the Production Manager of a medium-sized private enterprise, the challenge loomed large. Yet, he was determined to navigate the turbulence and land the company's ambitious goal.

Rajesh had decades of experience in precision manufacturing, but aerospace was a different ball game altogether. The Chief Operating Officer, Arvind Sharma, an accomplished veteran from the automotive sector, understood ISO 9001 well but was less familiar with the intricate requirements of AS9100D. Eager to meet the deadline, he initiated preliminary discussions with the team and laid out a framework, unknowingly overlooking several critical technological and procedural requirements unique to manufacturing aerospace parts and assemblies.

### The Audit Wake-Up Call

As the stage 2 audit began, the air in the factory felt heavy with anticipation. The lead auditor, Anand Raghavan, a seasoned professional with years of aerospace expertise, meticulously examined the processes. His findings hit like a cold wave: a nonconformance under clause 9.2.1.

*"Formal internal audits are conducted, but process deficiencies persist," he noted. "The effectiveness of processes is not evaluated, and customer, statutory, and regulatory requirements are not identified during the internal audit of the QMS. The organization is not serious enough about ensuring process conformity and effectiveness."*

The audit team found that issues flagged in internal audits remained unresolved, and corrective actions had been only corrections at the surface level. No systematic corrective actions were implemented. "If internal audits had been rigorous and NCs addressed proactively, the certification audit would have gone smoother," Raghavan remarked.

### Turning the Tide

The feedback was a jolt. Rajesh convened his team that evening, the factory's fluorescent lights flickering as they pondered over the audit findings. He reached out to a freelance lead auditor, Priya Menon, who agreed to guide them through the universal five-part corrective actions.

Priya's approach was methodical. She helped the team break down each NC, identify the temporary rectifications, detailed RCA, and implement corrective actions. Process effectiveness (KPIs) indicators were defined for each process, and rigorous internal audits became the norm. They integrated customer, statutory, and regulatory requirements into the QMS, ensuring every gap was closed. The audit program for the re-audit included process approach, KPI measurements, and checking conformity to the process requirements.

For weeks, the factory buzzed with focused energy. Production operators, quality engineers, and supervisors collaborated like never before. Training sessions were held, and employees gained a deeper understanding of AS9100D requirements. Slowly but surely, the company transformed. They were ready to fill up the corrective actions in the NCR formats and invited the auditor to carry out a re-audit of the processes in the organization.

## Cleared for Take-Off

When the re-audit day arrived, Rajesh felt cautiously optimistic. The same auditor, Raghavan, returned—with a friendly demeanor, but neutral as he revisited each process. But this time, the results were different.

The corrective actions were effective, process improvements were sustained, and the organization demonstrated a clear commitment to quality. Raghavan, after reviewing the final checklist, smiled.

"The improvements are commendable," he said. "You've built a foundation for long-term success. I'm recommending your certification."

The AS9100D certificate arrived a few weeks later, and the division celebrated the achievement—not just for the plaque on the wall but for the resilience and teamwork that made it possible.

Rajesh stood by the production floor, watching components being assembled with newfound precision. The journey had been tough, but the destination was worth it. He addressed his organization later. He highlighted the **benefits realized** by the organization:

The certification was not just a badge of honor but a strategic asset. With AS9100D accreditation:

- The company gained access to high-value aerospace contracts.
- Process efficiency improved, reducing waste and rework.
- Customer confidence surged, leading to increased orders and stronger relationships.
- Employee morale and competency grew through continual training and participation.
- The organization built a culture of quality, positioning itself for sustained growth and innovation.

Their confidence to deliver products to customers without defects increased, flight to certification had finally taken off, and the sky was the limit.

## CS 125 [AS9100D Clause 9.2.2] Fixing NCs in Internal Audits: An Aeroengine MRO Story

A large aeroengine MRO unit was undergoing its AS9100D certification audit. The auditor's keen eye quickly identified gaps in the internal audit processes. In his report, he documented multiple NCs under clause 9.2.2, highlighting the weaknesses in how internal audits were conducted.

Here is how each NC was addressed:

*NC (a): Audit program does not consider the importance of the processes.*

### Observation

The internal audit program failed to prioritize critical processes, such as quality control and safety-critical operations, over less impactful areas.

### Follow-Up Actions

1. **Risk-Based Prioritization:** The audit program was revised to incorporate a risk-based approach. Processes where there were many NCs last time and processes with higher importance or risks, such as engine teardown and assembly, were scheduled for more frequent audits.
2. **Process Mapping:** A detailed mapping of all organizational processes was undertaken to identify their relative importance.
3. **Training on Risk-Based Thinking:** Internal auditors were trained on how to align audit planning with process criticality.

*NC (b): Changes and results of previous audits not considered.*

### Observation

Internal audits did not account for organizational changes, such as new technology introductions, or address findings from previous audits.

### Follow-Up Actions

1. **Dynamic Audit Planning:** The audit program was updated to include a review of organizational changes and past audit results.
2. **Audit History Review:** All previous audits were revisited to identify recurring issues and ensure they were addressed.
3. **Change Management Integration:** Processes affecting internal audits were linked to change management protocols to automatically flag areas needing attention.

*NC (c): Internal audits not carried out by trained internal auditors (employees).*

### Observation

Untrained personnel were conducting internal audits, leading to inadequate assessments and missed issues.

### Follow-Up Actions

1. **Comprehensive Auditor Training:** All internal auditors were enrolled in certified training programs covering AS9100D requirements, audit techniques, and RCA.
2. **Competency Assessment:** A competency matrix was established to track auditor qualifications and ensure that only trained individuals conducted audits.

3. **Audit Shadowing:** New auditors shadowed experienced personnel before conducting independent audits.

*NC (d): No trend analysis to show sustained/ improved performance.*

### Observation

The organization lacked a system to analyze trends from internal audits, missing opportunities to track sustained or improved performance.

### Follow-Up Actions

1. **Trend Analysis System:** A trend analysis module was added to the audit management system, capturing recurring issues, resolved findings, and overall compliance levels.
2. **KPI Integration:** KPIs were defined for audit performance, such as the number of findings, time taken for resolution, and recurrence rates.
3. **Data Visualization:** Regular trend analysis reports with visual charts and graphs were presented during management reviews.

*NC (e): PEAR evaluation not conducted for all production processes.*

### Observation

PEARs were missing for several production processes, including engine testing and calibration.

### Follow-Up Actions

1. **PEAR Implementation Plan:** A schedule was created to ensure all production processes underwent PEAR evaluation (AS9101E can give you details).
2. **Standardized Templates:** Templates for PEAR evaluations were developed to streamline data collection and reporting.
3. **Training on PEAR:** Employees responsible for production processes were trained to carry out PEAR evaluations effectively.

*NC (f): Identified NCs remain open for over three months.*

### Observation

NCs identified during internal audits were not being closed promptly, causing delays in corrective actions.

### Follow-Up Actions

1. **NC Tracker Implementation:** An NC tracking system was deployed to monitor all findings, with automated reminders for pending actions.
2. **Closure Timelines:** A policy was established requiring all NCs to be closed within 30 days unless a valid extension was approved.
3. **Accountability Framework:** Owners were assigned for each NC, and closure status was reviewed during management meetings.

### Results and Follow-Up Audit

The corrective actions were implemented over a month, and the evidence of the updated processes was documented in the OASIS. During the follow-up audit, the auditor reviewed the following changes:

- The audit program now prioritized critical processes and incorporated risk-based thinking.
- Organizational changes and past audit results were reflected in the revised audit plans.
- Only trained personnel conducted audits, and their competency was documented.
- Trend analysis provided actionable insights, with clear evidence of sustained performance improvements.
- PEAR evaluations were systematically carried out for all production processes.
- All NCs were closed within the stipulated timelines, with clear accountability.

The auditor commended the MRO unit for addressing the gaps comprehensively and emphasized the importance of sustaining these improvements.

As the team celebrated their progress, the auditor encouraged them: "Internal audits are the backbone of your QMS. When done right, they ensure compliance and drive excellence."

The organization embraced this philosophy, embedding internal audit improvements into its culture of continual improvement.

## CS 126 [AS9100D Clause 9.2.2] Evaluate Readiness for Audits of QMS

During the certification audit for AS9100D, the sub-assembly manufacturing organization faced a significant challenge when the auditor uncovered a glaring NC in their management review processes.

The auditor highlighted with evidence that **clause 9.3.2** was not being effectively implemented:

*Management review/divisional quality review did not address mandatory agenda points, including:*

*(a) OTD performance*

*(b) Defects reported by customers*

*(c) Performance of suppliers*

*(d) Effectiveness of actions taken to address organizational risks and opportunities*

*(e) Changes in external and internal issues relevant to the QMS/business*

Management Representative Ajay Menon, acknowledged the gaps, explaining that reviews had primarily focused on production output and immediate concerns, neglecting these critical elements.

The auditor emphasized the importance of these agenda points in driving improvement, risk mitigation, and maintaining customer satisfaction. He requested Menon to initiate follow-up actions for each point, including containment, RCA, and implementation of corrective actions.

## Addressing the Agenda Points

### (a) On-Time Delivery Performance

**Observation:** No formal review or trend analysis of OTD performance was conducted during management review meetings.

**Follow-Up Actions:**

1. **Data Collection:** Historical OTD data were consolidated and reviewed to identify patterns.
2. **RCA:** Delays were traced to specific issues, such as supply chain disruptions and inefficiencies in production planning.
3. **Corrective Actions:** The organization implemented better production scheduling tools and introduced penalties for late supplier deliveries.
4. **Performance Monitoring:** OTD performance was added as a permanent agenda item in all management review meetings, with monthly tracking against targets.

### (b) Defects Reported by Customers

**Observation:** Customer complaints and defects were not systematically reviewed during management review meetings.

**Follow-Up Actions:**

1. **Complaint Database:** All customer-reported defects were logged and categorized.
2. **Trend Analysis:** Patterns of recurring issues were identified, particularly related to assembly tolerances.
3. **Corrective Actions:** Assembly line processes were refined, and additional inspections were introduced at critical stages.
4. **Customer Feedback Review:** Defect analysis and customer feedback became a fixed agenda item in management review meetings to ensure continual improvement.

### (c) Performance of Suppliers

**Observation:** Supplier performance, including quality and delivery metrics, was not regularly reviewed.

**Follow-Up Actions:**

1. **Supplier Scorecards:** A supplier performance evaluation system was introduced, covering quality, delivery, and responsiveness.
2. **Supplier Engagement:** Low-performing suppliers were identified, and improvement plans were implemented in collaboration with them.
3. **Periodic Review:** Supplier performance data were made a standing management review agenda item, ensuring consistent monitoring and follow-up.

### (d) Effectiveness of Actions Taken to Address Organizational Risks and Opportunities

**Observation:** No evidence showed that actions addressing risks and opportunities were reviewed for effectiveness.

**Follow-Up Actions:**

1. **Risk and Opportunity Registry:** A centralized registry was created to track identified risks, opportunities, mitigation actions, and outcomes.
2. **Effectiveness Metrics:** Metrics were defined to assess the success of mitigation strategies, such as reduced downtime or improved production efficiency.
3. **Action Reviews:** The effectiveness of actions addressing risks and opportunities was added to the management review agenda, with outcomes discussed and documented.

### (e) Changes in External and Internal Issues Relevant to the QMS/Business

**Observation:** The organization failed to consider changes, such as new customer requirements or industry regulations, during management review meetings.

**Follow-Up Actions:**

1. **Change Monitoring Framework:** A structured process was established to monitor changes in external and internal issues, including market trends, regulatory updates, and business expansions.
2. **Integration into Management Review:** A dedicated session on these changes was added to the management review agenda to assess their impact on the QMS and define necessary actions.
3. **Stakeholder Inputs:** Inputs from key stakeholders, such as department heads and regulatory experts, were incorporated into the review process.

## Follow-Up Audit

A month later, the auditor returned to verify the corrective actions. The organization presented:

- Updated management review meeting minutes addressing all five agenda points with detailed discussions and decisions.
- Evidence of OTD performance tracking, customer defect trend analysis, and supplier scorecards.
- Documented reviews of risk mitigation effectiveness and changes affecting the QMS.
- Training records showing management review participants were educated on the updated process.

The auditor expressed his satisfaction with the actions taken and noted the evidence on OASIS. "You've addressed the gaps comprehensively, and your management review process now aligns with AS9100D requirements," he remarked.

However, he reminded the team, "The true measure of success is consistency. During the surveillance audit, I will verify whether these improvements are sustained and whether they deliver tangible results."

## A Stronger Foundation for the Future

The experience was a turning point for the organization. By embracing a structured and comprehensive management review process, they strengthened their QMS and improved their ability to meet customer and regulatory expectations.

As the auditor departed, he left a piece of advice: "Management review is not just about compliance—it is your opportunity to steer the organization toward excellence."

The organization took this to heart, ensuring that management review meetings became a strategic tool for continual improvement and sustained success.

Chapter 10

# Nonconformities in Improvement Processes

## CS 127 [AS9100D Clause 10.2.1] Audit Debate: Addressing Human Factors in Defect/ Accident Investigations

A routine AS9100D audit was underway at the headquarters of AeroDynamics Ltd., a prominent aircraft manufacturer. The audit proceeded smoothly until the discussion shifted to clause 10.2.1(b) of the standard: the requirement to address human factors when investigating NCs, including customer complaints.

The auditor, Kavita Mehta, began by reviewing defect and accident investigation reports. A pattern emerged: several reports cited "human error" as the root cause, with no further action taken to address systemic contributors.

### The Initial Discussion

**Auditor:** "I notice several instances where defect investigations conclude with 'human error' as the root cause. Can you explain the rationale behind these conclusions?"

**Quality Manager:** "In many cases, it is clear. For instance, if a technician uses the wrong torque setting despite clear instructions, it is obviously human error."

**Auditor:** "And what steps do you take to ensure this type of error doesn't recur?"

**Quality Manager:** "We retrain the individual or issue a reminder about procedures. Beyond that, it is difficult to address what seems to be an isolated lapse in judgment."

Mehta paused, her expression neutral but focused. "Let's discuss a specific case," she said, pulling up a report on a recent accident involving a pilot.

### The Case of the Fatal Accident

The report described an incident where a pilot failed to deploy the landing gear, resulting in a crash. The conclusion stated, "Pilot error: failure to follow checklist procedures. No further action is required. The pilot has already paid the penalty with his life."

Mehta's eyebrows rose slightly as she read this aloud. The room grew tense.

**Auditor:** "This conclusion raises concerns. Assigning blame to a deceased pilot might seem sufficient, but clause 10.2.1(b) requires organizations to explore deeper causes. Why wasn't a systemic analysis conducted?"

**Head of Safety:** "With due respect, what more can we do? The pilot missed a step; it is tragic, but it is a clear case of human error."

**Auditor:** "Human error rarely exists in isolation. Were there factors such as fatigue, inadequate training, or cockpit design flaws that contributed to this mistake?"

The room buzzed with murmurs.

### The Debate on Human Factors

**Head of Engineering:** "I agree that 'pilot error' can be reductive, but is not it sometimes accurate? Overcomplicating investigations won't bring the pilot back."

**Auditor:** "True, but the goal is not just assigning fault. It is about prevention. Modern aviation has moved toward understanding *why* errors occur—whether they stem from poor interface design, unclear procedures, or systemic gaps."

**Head of Operations:** "But resources are finite. Investigating every angle for every defect is not feasible."

**Auditor:** "It is not about investigating every angle but the *right* ones. Take cockpit design: Was the landing gear control easily accessible? Were alarms clear and timely? Ignoring such factors risks repeating the same tragedy."

**Quality Manager:** "But what about cases where there genuinely is no systemic issue—just a lapse?"

**Auditor:** "Even then, consider mitigations like fatigue management programs, enhanced checklists, or peer reviews to minimize lapses. Simply labelling it as 'human error' without exploring these options doesn't align with the requirements of the standard AS9100D."

### The NC Issued

*After hours of discussion, Ms. Mehta issued the NC: vide clause 10.2.1 (b) "When an NC occurs, including any customer complaints, the organization shall determine the causes of the NC, including, as applicable, those related to human factors. This has not consistently occurred in defect/accident investigations, particularly in cases attributed to human error or one-time occurrences. No systemic corrective actions have been implemented to eliminate these causes permanently."*

### Post-Audit Actions

The audit led to significant introspection at AeroDynamics Ltd. The leadership team resolved to:

1. **Revise Investigation Procedures:** Mandate that all defect investigations include analysis of systemic contributors to human error.
2. **Establish a Human Factors Board:** Create a multidisciplinary team to review incidents with potential human factors involvement.
3. **Enhance Training:** Provide investigators with training in human factors analysis and root cause methodologies.
4. **Invest in Technology:** Explore tools like fatigue monitoring systems and enhanced simulation-based training.

### Conclusion

The NC was closed by the auditor after verification of corrective actions. Auditor said that effectiveness of corrective actions will be verified during the next surveillance audit. The organization received their AS9100D certificate subsequently.

The audit served as a turning point, shifting AeroDynamics Ltd.'s approach from blame assignment to a systemic analysis culture. The organization embraced the philosophy that understanding human factors is not about pointing fingers—it is about building safer systems and ensuring that similar incidents do not occur in the future.

## CS 128 [AS9100D Clause 10.2.1] Audit Findings: Receiving Inspection Area

During a routine audit at the receiving inspection area, the auditor noticed a concerning situation related to the handling of a nonconforming product. After issuing an NCR, the auditor narrated the following sequence of events to clarify the context without compromising the impartiality of the NCR:

A nonconforming product, supplied by an external provider, had been identified during receiving inspection. Instead of following the organization's established procedures for handling nonconforming products, the production manager intervened and directed the nonconforming product to be taken inside the firm for rework. This action was taken without issuing a recall notice or tagging the product for traceability.

The supplier was informed of the NC and requested to replace the defective item. However, no further corrective or preventive actions were communicated to the supplier, as required under clause 10.2.1(C) of the quality management standard. This clause mandates that when an NC occurs, including those arising from supplier-related issues, the organization must flow down corrective action requirements to the external provider responsible for the NC.

The inspector mentioned that this lapse occurred because the item was urgently required to meet delivery commitments for a final product to the customer. Despite the urgency, this deviation from procedure posed a significant compliance risk, as there was no documented recall process to ensure that defective parts could be replaced or reworked properly before the final product was delivered to the customer.

The auditor also noted that the production manager appeared unaware of the ongoing audit, suggesting a gap in communication and preparedness for such events.

### Auditor's Recommendations for Corrective Actions

The auditor identified several critical follow-up actions to address the observed lapses:

1. RCA:
    - Conduct a thorough investigation to determine why established procedures for handling nonconforming products were not followed.
    - Assess whether the production manager's actions were due to insufficient training, communication gaps, or pressure to meet delivery deadlines.
2. Immediate Corrective Actions:
    - Ensure the supplier is formally notified of the NC and required corrective actions are documented and communicated.
    - Implement a procedure to track and verify supplier responses to NC notifications.
3. Process Improvements:
    - Reinforce the organization's nonconforming product handling process to ensure that recall notices or equivalent documentation are issued whenever a defective product is identified.
    - Enhance the traceability system to prevent nonconforming products from bypassing established procedures.
4. Training and Awareness:
    - Provide targeted training to the production manager and other relevant personnel on NC handling requirements and compliance expectations.
    - Increase the awareness of the importance of adhering to procedures, even during high-pressure situations.
5. Verification of Effectiveness:
    - Schedule a follow-up audit to confirm that corrective actions have been implemented effectively and that compliance with clause 10.2.1(C) is being maintained.

### Next Steps

The auditor emphasized the importance of revisiting the area to validate the effectiveness of the corrective actions once they are implemented. This revisit will include:

- Reviewing updated documentation and records for compliance with nonconforming product handling procedures.
- Verifying that training has been conducted and that all relevant personnel are aware of their roles and responsibilities.
- Inspecting the process flow to ensure that corrective action requirements are properly communicated to external suppliers and followed up for closure.

Through these steps, the organization can strengthen its compliance framework, mitigate risks, and improve overall process reliability.

## CS 129 [AS9100D Clause 10.2.1] Audit Story: Updating Risks and Opportunities with Defect Reports

Skyward Aviation Systems, a manufacturer of precision components for military aircraft, was undergoing an audit for AS9100D certification. The audit focused on compliance with quality management standards, including the requirement to address NCs and update associated risks and opportunities.

### The Audit Observation

During the review of NCRs and defect investigation documents, the auditor, Ravi Kumar, noticed a recurring omission: none of the reviewed reports mentioned whether risks and opportunities identified during planning had been updated in response to the NC.

**Auditor:** "I've reviewed 20 NCRs from the past year, and while you've detailed root causes and corrective actions, there's no indication of updates to the risks and opportunities identified in your planning process. Can you explain why this step has been omitted?"

**Quality Manager:** "We prioritize fixing the issues and preventing recurrence. Updating risks and opportunities hasn't been part of our process because we feel that addressing the immediate issue is sufficient."

**Auditor:** "AS9100D clause 10.2.1(b) explicitly requires organizations to update risks and opportunities during NC handling. Without this, you risk recurring issues or missing potential improvements."

The quality manager looked uneasy, and the room fell silent.

### A Specific Example

Management Representative Kumar pointed to a recent NCR:

- Issue: A batch of titanium components failed during stress testing.
- Root Cause: Improper heat treatment due to an undocumented process deviation.
- Corrective Action: Retrain staff and add a checklist for the heat treatment process.

**Auditor:** "This was a significant issue that could impact customer trust and delivery timelines. Did your team update the risk register or analyze opportunities to improve overall process control?"

**Production Head:** "We didn't consider that necessary. The corrective action addressed the problem, and the risk register wasn't reviewed."

**Auditor:** "And if a similar deviation occurs in another process? Failing to link NCRs with your risk management process means you're not proactively mitigating future risks or identifying systemic OFI."

### The NC Issued

After reviewing the evidence, Kumar issued an NC:

---

*NC statement: "When an NC occurs, including any customer complaints, the organization shall determine the causes of the NC and update risks and opportunities determined during planning, if necessary. This step has been consistently omitted in reviewed NCRs and defect reports, posing a risk to the organization's ability to manage future NCs proactively."*

---

## Follow-Up Actions for the NC

The leadership team of Skyward Aviation Systems convened immediately after the audit to address the NC.

### Immediate Containment Actions

- Conduct a review of all NCRs from the past 12 months to identify cases where the register for risks and opportunities should have been updated.
- Document the missed updates and propose revisions to the risk register.

### RCA

- Use tools like the 5 Whys method to determine why this requirement was overlooked.
- Identify gaps in awareness of AS9100D requirements and the linkage between NCRs and risk management.

### Corrective Actions

- Update the NC handling procedure to include a mandatory step for reviewing and updating risks and opportunities.
- Modify the NCR template to include a section for documenting changes to risks and opportunities.
- Ensure that defect investigation reports include a clear reference to how risks and opportunities were reassessed.

### Training and Awareness

- Conduct training sessions for all quality management staff to emphasize the importance of integrating risk management into NCR handling.
- Include examples of how to identify and document updates to risks and opportunities effectively.

### Verification and Monitoring

- Implement a periodic internal audit to verify that all NCRs include updates to risks and opportunities as required.
- Assign a quality lead to review each NCR for compliance with this requirement before closure.

### Preventive Measures (Risk Mitigations)

- Integrate risk and opportunity updates into the organization's ERP system to ensure automated prompts during NCR processing.
- Create a dashboard for management to monitor updates to risks and opportunities in real time.

### Outcome and Lessons Learned

Within three months, Skyward Aviation Systems completed the corrective actions and provided the auditor with evidence of compliance. The quality manager reflected on the experience:

**Quality Manager:** "This NC has been a wake-up call. By linking NCRs to our risk management process, we've gained a clearer picture of our vulnerabilities and a pathway to continual improvement."

**Auditor:** "That's the purpose of AS9100D—not just fixing problems but preventing the problems from happening again and finding ways to continually improve."

The organization not only resolved the NC but also strengthened its QMS, aligning better with the principles of proactive risk management and continual improvement.

## CS 130 [AS9100D Clause 10.2.1] Unserviceable Part: A Certification Audit Tale

It was a bright, sunny day in Saudi Arabia when the aircraft equipment division geared up for the all-important certification audit. The management team was determined to secure the prestigious certification, and the process owners were brimming with confidence, clutching their meticulously prepared procedures. But as fate would have it, not everything was in perfect harmony.

At the receiving inspection area, the auditor, a seasoned detective of NCs, was quietly reviewing inspection records retained for the past three months. The atmosphere was calm, almost too calm when suddenly a glaring oversight emerged: an *unserviceable part*.

### The Unfolding Drama

Here's the twist: A month earlier, the inspection team had flagged a defective part supplied by an external provider. The inspector, ever vigilant, immediately informed the purchase manager and suggested withholding payment for the faulty item. So far, so good.

But then, the script went off track. Instead of activating the supplier corrective action request (SCAR) procedure—an established process designed to handle such issues—the purchase manager decided to let the matter quietly slip into obscurity. No SCAR was issued. No corrective actions were requested from the supplier. The defective part simply lingered, a silent witness to procedural neglect.

When the auditor returned a month later for a follow-up, he found the situation unchanged. No action had been taken. The defective part sat there, and the SCAR process, like an unused superhero cape, remained untouched.

### The Auditor's Assessment

The auditor, unimpressed but professional, issued an NCR under **clause 10.2.1(d)** major:

> *Flow down of corrective action requirements was not evidenced when an external provider supplied an unserviceable part. Although a SCAR procedure exists in the purchase process, no evidence of its implementation was found. Immediate action is required to address this gap.*

### The Management Response

Management, eager to protect their certification dreams, promised swift and systematic corrective actions. At a high-energy meeting, they laid out their plan with clarity and determination:

1. **SCAR Takes Centerstage:** From now on, every supplier-related NC will be met with a SCAR. Suppliers will be promptly informed of issues and required to provide corrective actions, while rejecting their supply.
2. **Training for Everyone:** A mandatory crash course on SCAR procedures for all relevant personnel. Even the coffee machine might get a lesson in compliance. (*"Excellence in a cup begins with compliance in the process. Adhering to regulations is not just about following rules; it is about crafting the finest coffee experience." Refer* https://primosltd.com/uk-commercial-coffee-machine-compliance-guide/#article.)
3. **Tracking for Success:** A new system to monitor supplier responses and ensure follow-up actions are completed on time. No more loose ends.

4. **Communication to Suppliers:** A formal reminder to all suppliers: defective parts are not welcome here. If issues arise, expect a SCAR and a request for corrective action faster than a jet engine can rev up. If a counterfeit part is detected during receiving inspection, the same will be confiscated and legal action will be initiated, and the concerned government enforcement authorities will be informed (reference: 2).

### A Friendly Word to Suppliers

To all our valued suppliers:

We appreciate working with you, but we need your help to keep quality flying high. Please ensure all parts meet our standards before delivery. If there's a hiccup, we'll work with you to resolve it through the SCAR process. Together, we can maintain excellence and aim for the skies—certified, compliant, and defect-free!

### The Lesson Learned

This incident served as a wake-up call. The unserviceable part may have been a small oversight, but it highlighted a bigger gap in the system. With the corrective actions now underway, the aircraft equipment division is back on course, ready to ace its certification audit and set new standards in quality and compliance.

Let us ensure this tale becomes a thing of the past and focus on building a future where every part, process, and procedure aligns perfectly with our commitment to excellence.

## CS 131 [AS9100D Clause 10.3] The Kaizen Detective: Who Forgot to Water the Garden?

In the Aircraft equipment division in Saudi Arabia, the air was thick with determination. The team was in the middle of their certification audit, and everyone—from management to process owners—was laser-focused on making a strong impression.

On this day, the auditor, a seasoned expert with a knack for uncovering gaps, was stationed at the management representative's office. The conversation started amicably, with the auditor exploring the organization's improvement initiatives beyond corrective actions. The management representative enthusiastically shared details of various Kaizen activities that had been launched over the past few months, aimed at boosting efficiency and optimizing processes.

### A Closer Look at Kaizen

The auditor, intrigued by the enthusiasm, asked to see records of these improvement activities. He reviewed the documented Kaizen projects initiated in the last three months, noting the ambitious goals and plans outlined for each.

However, as he delved deeper, a glaring issue emerged:

- While the Kaizen activities were well documented in their initiation phase, there were **no records of monitoring their implementation**.
- Even more concerning, there was **no evidence of evaluating the effectiveness of these initiatives** to determine whether they achieved their intended results.

It was like planting a garden but never checking to see whether the flowers bloomed.

### The Auditor's NC

The auditor issued an NC under **clause 10.3**:

> *The organization initiated some Kaizen activities, but there is no record of monitoring the implementation of improvement activities and evaluation of the effectiveness of the results.*

This NC was a wake-up call. The lack of follow-through on improvement initiatives undermined the organization's commitment to continual improvement—a cornerstone of the certification process.

### Management's Commitment

The management representative, visibly concerned but determined to address the issue, promised the auditor a prompt and systematic response. A plan was quickly drafted to close the gap and ensure future compliance:

1. **Kaizen Monitoring Framework:** A dedicated tracking system would be implemented to monitor the progress of all Kaizen activities. This system would include timelines, milestones, and assigned responsibilities.
2. **Effectiveness Evaluation:** Regular review sessions would be scheduled to evaluate the outcomes of improvement initiatives. Metrics and KPIs would be established to measure success objectively.
3. **Training on Continual Improvement:** Training programs would be rolled out to educate all relevant personnel on the importance of monitoring and evaluating improvement activities.
4. **Integration with QMS:** Kaizen activities would be fully integrated into the QMS, ensuring they are treated with the same rigor as corrective actions.

### A Lesson in Follow-Through

As the meeting concluded, the management representative reflected on the auditor's feedback. It was not enough to start improvement projects; the real value came from seeing them through and learning from their outcomes.

With a renewed sense of purpose, the team got to work, determined to turn this NC into a success story. They committed to not just meeting certification standards but embracing continual improvement as a way of life.

### The Road Ahead

The aircraft equipment division realized that Kaizen is not just about ideas—it is about results. By implementing the auditor's recommendations, they aimed to build a culture where every improvement initiative was tracked, measured, and celebrated for its impact.

And so, the Kaizen garden began to bloom, nurtured by diligent monitoring and a commitment to excellence. Certification was no longer just a goal; it was a testament to their dedication to continual growth.

Chapter 11

# Action for Indicated Opportunities for Improvement

Audit findings include OFI. During audits, when processes are found conforming and no NCs are identified, it is a testament to the organization's commitment to maintaining quality standards. However, the role of a highly experienced auditor goes beyond mere compliance verification. To truly elevate the organization toward business excellence, auditors often identify and indicate OFI. This is considered a value-adding activity during audits.

These OFIs do not require mandatory follow-up actions, but they are valuable insights aimed at unlocking untapped potential within the organization. They serve as a guide for enhancing efficiency, strengthening systems, and fostering innovation. While organizations may choose whether to act on these suggestions, those that approach OFIs with diligence and commitment often reap significant benefits—improved processes, increased customer satisfaction, and stronger competitive edge.

An auditor's passion for quality excellence shines through when they identify these opportunities. The intention is not just to find faults but to contribute to the organization's growth and success. By adopting these suggestions with enthusiasm, organizations can demonstrate their willingness to go beyond compliance, embracing continual improvement as a core value.

Taking proactive action on OFIs reflects a forward-thinking mindset, one that seeks not only to meet but to exceed expectations. It signals to customers, employees, and stakeholders that the organization is dedicated to achieving and sustaining the highest levels of performance.

As such, OFIs should not be viewed as mere observations but as stepping stones toward a future of excellence. The encouragement and insights provided by auditors are a gift—one that, when embraced, can transform the ordinary into the extraordinary.

## CS 132 [AS9100D Clause 8.4.1] Strengthen Supplier Evaluation Metrics

Aerotech Innovations, a mid-sized aerospace component manufacturer certified to AS9100D, was buzzing with energy as the audit commenced. Their consistent performance in producing lightweight, durable components has earned them accolades in the aerospace sector. Yet, like every organization striving for excellence, they welcomed the opportunity to refine their processes.

Dr. Alan Mercer, a seasoned auditor with an unwavering passion for quality, walked into the final audit meeting with a warm smile. "First of all," he began, "I must commend your team for a truly robust QMS. The dedication of everyone here to maintaining compliance with AS9100D is evident and commendable. Your processes are solid, but I see a spark of potential to take them even further."

As the team listened intently, Dr. Mercer pointed out an opportunity: improving supplier evaluation by adopting predictive analytics. "By analyzing historical data trends and creating predictive models, you can foresee and prevent disruptions before they occur. Imagine a world where potential delays never impact your production timelines—where you are not just reacting to issues but staying two steps ahead."

The Quality Manager, Emma Reyes, nodded thoughtfully. "This is brilliant. We've been focusing so much on reactive measures that we've overlooked the proactive side. Thank you for this insight—it could really transform our supplier relationships."

Dr. Mercer's voice softened as he added, "You're already doing an exceptional job, Emma. This is not about fixing a problem but about unlocking a new level of excellence. Imagine the confidence your customers will feel knowing your supply chain is bulletproof."

The team felt inspired and motivated to act on the suggestion. As they expressed their gratitude, the MD remarked, "This is why audits are so valuable. Thank you, Dr. Mercer, for reminding us of the power of continual improvement. We are excited to implement this!"

## CS 133 [AS9100D Clause 7.2] Optimize Training Programs with Real-Time Performance Feedback

Orion Defense Systems, a cutting-edge avionics producer, had gathered its leadership team for the closing meeting of their audit. The atmosphere was hopeful yet slightly tense—audits always came with the promise of improvement, but also the fear of lapses.

Dr. Mercer, known for his meticulous yet encouraging approach, began with a heartfelt acknowledgement. "First, let me say that the pride you take in your work is unmistakable. I've visited countless organizations, but few match the commitment I've seen in your training and quality teams. Your dedication to excellence is not just ticking boxes; it is a way of life here."

Smiles began to spread around the room as Dr. Mercer continued, "That said, even great systems can shine brighter. Your training programs are exceptional, but integrating real-time performance feedback could bring about a significant transformation. Imagine a technician immediately understanding and correcting their errors as they train. It is a simple step with monumental outcomes—better skills, fewer mistakes, and, ultimately, happier customers."

Liam Cheng, the Head of Quality, leaned forward, his excitement palpable. "That's an incredible idea. We've been exploring digital tools, but this focus on instant feedback adds a whole new dimension. It feels like the missing piece."

Dr. Mercer nodded, his tone encouraging. "You've already built the foundation—this is just the next logical step. And knowing the passion of this team, I have no doubt you'll make it happen beautifully."

As the meeting wrapped up, the CEO stood to express the team's appreciation. "Dr. Mercer, your words have energized us. This is not just an OFI; it is an opportunity for growth. Thank you for showing us what's possible."

## CS 134 [AS9100D Clause 7.5.3] Enhance Document Control with Automated Workflows

Zenith Aerospace Technologies, a key player in precision engineering, had just concluded the week-long audit. Their processes had passed with flying colors, but the team eagerly awaited feedback from Dr. Mercer, known for offering insights that pushed companies to new heights.

As he stood before the leadership team, Dr. Mercer began with sincere admiration. "Let me start by saying this: Zenith Aerospace is a name synonymous with quality. Your commitment to precision is not only evident in your products but in every layer of your QMS. It is truly inspiring to see."

The team's pride was evident, but they leaned in as Dr. Mercer shifted gears. "Your document control processes are compliant and effective. However, I see an opportunity to make them even more efficient by automating workflows. Imagine a system where approvals happen in real time, where no updates are missed, and your team can focus more on innovation than paperwork."

Sarah Patel, the Operations Manager, was immediately intrigued. "Automation has been on our radar, but we weren't sure where to start. This suggestion feels so actionable—and it could save us countless hours."

Dr. Mercer smiled warmly. "Sarah, your team is already doing amazing work. Think of this as a way to make their lives easier and your operations even more seamless. With your drive and expertise, I have no doubt you'll implement this flawlessly."

As the leadership team clapped in appreciation, the CEO addressed Dr. Mercer with gratitude. "Your feedback has been nothing short of transformative. Thank you for helping us see opportunities we might have overlooked. We'll act on this—and when we succeed, you'll have played a big role in it."

Each organization left the audit not just with compliance but with renewed determination to achieve excellence. Dr. Mercer's insights reminded them that the journey of quality is not about being "good enough"—it is about striving for greatness every single day.

## CS 135 [AS9100D Clause 8.2] Fostering Continual Improvement in Engineering Services

The auditor commended your organization's D&D process, stating that it is excellent. Customers highly value your approach to optimizing designs and facilitating ease of manufacture, with minimal defects or deviations during production.

Your team is already proactive in identifying key characteristics, critical items, interchangeable components, and special process requirements. However, the auditor suggested a simple yet impactful improvement:

To enhance control during manufacturing, consider documenting special requirements, critical items, and key characteristics in a common table within the design outputs. This will make it easier for teams to quickly reference and manage these elements throughout production.

### Key Insights

- Interrelation of Elements: Special requirements, critical items, and key characteristics are closely connected—one can lead to the identification of another.
- Special Requirements (clauses 8.2.2 and 8.2.3): Identified during requirement determination and review processes.
- Critical Items (clause 8.3.5): These may arise from special requirements and require specific actions to ensure proper management.
- Key Characteristics: A subset of critical items where variation must be tightly controlled to maintain product quality and performance.

### Recommendation

Make it a practice to systematically identify and document these elements in every design output. This effort will significantly support the manufacturing team, ensuring critical factors are well managed and the final product consistently meets quality standards.

## CS 136 [AS9100D Clause 10.3] Strengthening Continual Improvement in Engineering Services: Closing Remarks 1

Good afternoon, everyone. It has been a pleasure conducting this audit and interacting with your talented team. I want to take a moment to commend your organization for the excellent way you deliver engineering services. Your reputation for quality and innovation is well known, and it is clear why—your commitment to excellence is reflected in the processes I have observed.

Even the most successful organizations can find ways to enhance their systems, and AS9100D offers valuable guidance for continual improvement. Let me share an opportunity for growth that can help you further refine your services.

While planning for engineering services, it is essential to identify clear criteria for the acceptance of those services. I'm pleased to see that you already have this in place, and I'm especially impressed by your extensive use of automation to streamline your processes. This is a significant strength that positions you for future success.

However, given the nature of your products and the unique characteristics of engineering services—where physical measurements may not always apply—there is an opportunity to establish simpler yet effective acceptance criteria. In those areas, I encourage you to explore the use of statistical techniques for process optimization, such as:

- Process control to monitor and maintain stability.
- Selection and verification of key characteristics to focus on critical service factors.
- Process capability measurements to understand and enhance performance limits.
- Statistical process control and verification to drive data-based decision-making and continual improvement.

By thoughtfully evaluating and applying these techniques, you can further strengthen the consistency and reliability of your services, reinforcing your industry leadership.

### After the Organization's Response

I am pleased to note that your organization has responded positively to this feedback, expressing enthusiasm for exploring the suggested improvements. Your willingness to embrace new approaches and your openness to continual learning are qualities that will undoubtedly drive your future success.

I understand that you are eager to evaluate these recommendations and consider their practical application within your existing framework. This proactive mindset will not only enhance your internal processes but also further solidify your standing as a trusted engineering services provider for the aerospace and defense industry.

Of course, the decision is entirely yours, and I have no doubt that you will carefully consider this recommendation. I am confident that with your dedication to quality and innovation, your organization will continue to thrive and set new benchmarks in the engineering services sector.

Thank you for your cooperation and hospitality during the audit—I look forward to hearing about your future successes and perhaps even seeing your achievements recognized in industry publications or award functions!

## CS 137 [AS9100D Clause 10.3] Strengthening Continual Improvement in Engineering Services: Closing Remarks 2

Good afternoon, everyone. It has been a pleasure to conduct this audit and witness the dedication your organization demonstrates toward maintaining a robust QMS. I want to take a moment to acknowledge the commendable efforts you've made so far and share some insights on opportunities for further growth and excellence.

I observed that your **quality engineering team** conducts data analysis and evaluations, though this happens occasionally rather than consistently. Your **management team** actively drives production activities, which contributes to steady progress and the delivery of defect-free products. And overall, your **improvement processes** are compliant and effective—but I must tell you, there is immense potential to elevate these processes even further to thrive in today's competitive market.

In that spirit, let me share a few OFIs that could help propel your organization to greater heights:

1. **Frequent Monitoring and Evaluation:** Instead of conducting assessments occasionally, consider implementing regular, structured reviews of improvement activities. This ensures timely course corrections, sustains progress, and keeps your processes agile and responsive.
2. **Proactive and Innovative Improvement Initiatives:** Do not just address immediate issues—cultivate a proactive mindset that continually seeks out ways to optimize processes, enhance product quality, and boost overall performance. Let innovation be a constant driving force within your teams.
3. **Learning and Benchmarking Best Practices:** Tap into lessons learned from past experiences, customer feedback, and even industry leaders. Benchmarking against best practices will provide valuable insights and guide your continual improvement journey.
4. **Recognition and Motivation:** A culture of improvement flourishes when employees feel valued. Recognize and reward those who contribute to innovative solutions and process enhancements. Public celebrations of these achievements can inspire teams and reinforce a growth-oriented mindset. And who knows? Perhaps your organization's success will soon make headlines—with news of award functions honoring your most innovative minds!

### Looking Ahead

By identifying specific areas for targeted improvement, developing a structured plan with periodic evaluations, and fostering an organizational mindset of innovation and excellence, you position yourselves not just to keep up with the industry but to lead it.

I truly believe that your dedication to continual improvement will pay off, and I look forward to reading about your achievements in the future. Maybe I'll come across a newspaper article that reads: "Organization Honors Employees for Outstanding Innovative Contributions—A Celebration of Continual Improvement and Excellence."

And let me leave you with a light-hearted thought: "Improvement is like compound interest—the more you invest in it, the greater the rewards!"

I have no doubt that by embracing these opportunities, your organization is well on its way to setting a shining example of excellence for others to follow. Thank you!

# Index

## S



## T

# About the Author

## Thomas Kannattumadom

*Aerospace Industry Expert | Lead Auditor | Author | Guide | Trainer*

Courtesy of Thomas Kannattumadom.

## Profile Summary

Shri Thomas Kannattumadom is a distinguished quality professional in the aerospace sector with over four decades of experience in quality assurance, audit, and training. He is a **certified Lead Auditor for ISO 9001 and AS9100D**, registered with International Aerospace Quality Group (IAQG)–Online Aerospace Supplier Information System (OASIS) (OIN 6144704664) and approved by PROBITAS/IAQG valid until January 2022. He also holds the prestigious Aerospace Industry Experienced Auditor **(AIEA) certification (Cert. No. 1619032761).**

Over the past 16 years, Mr. Thomas has conducted more than **1000 man-days of aerospace quality audits** across India, Sri Lanka, the UAE, Singapore, Malaysia, Thailand, Saudi Arabia, Japan, and the UK for AS9100D/EN9100:2018 certifications. His commitment to excellence in the field has been acknowledged by the **Quality Council of India (QCI)**, which appointed him as a **SAMAR Assessor** in 2023.

He is a **visiting faculty member at HAL Management Academy, Bangalore**, and has delivered over **300 training days** to major organizations, including Defence Research and Development Organisation (DRDO) labs, HAL divisions, and other public and private aviation, space, and defense institutions.

## Career Background

Mr. Thomas superannuated on February 28, 2006, as Deputy Director/Principal Scientific Officer at the Directorate General of Aeronautical Quality Assurance (DGAQA), Ministry of Defence, with an illustrious career spanning nearly 40 years.

His experience includes

- 11+ years of teaching mechanical engineering subjects at the College of Military Engineering, Pune.
- 2 years in quality assurance of Engineering Equipment, at Directorate General of Quality Assurance (DGQA), Pune.

- 28 years in aeronautical quality assurance, overseeing the development, production, and overhaul of aircraft, aeroengines, and airborne systems for the Indian Air Force—from trainers to frontline fighters.

He was also a **technical member on various aircraft/engine defect and accident investigations** and contributed to special study and audit teams for the Indian Air Force and the Ministry of Defence. His efforts have had a lasting impact on **safety and quality in Indian military aviation**.

## Publications

Mr. Thomas is a seasoned author whose work reflects both technical expertise and personal reflection. His published books include

1. *Safety Follows Where Quality Leads*, February 2010, Orchard Publishers.
2. A Complete Guide to Implement Aerospace Standard AS 9100 C – Including ISO 9001:2008, AS 9110 Rev A, and AS 9120 Rev A, August 2012, Orchard Publishers.
3. Implement AS 9100 D/EN 9100:2018 for Business Excellence: Quality Management Systems - Requirements for Aviation, Space and Defence Organizations (Same as EN 9100: 2016; Includes ISO 9001:2015). August 2025, River Publishers, Denmark.

https://www.riverpublishers.com/book_details.php?book_id=1427

4. *Beginnings and Breakthroughs: My Journey to Midlife*, January 2025, Notion Publishers.
5. *Reflections and Revelations: The Later Years*, January 2025, Notion Publishers.

He has also presented technical papers at multiple national and international conferences and seminars.

www.ingramcontent.com/pod-product-compliance
Lightning Source LLC
LaVergne TN
LVHW070947210926
848482LV00002B/119
*9781468609660*